# NASA

A STORY OF MANNED

&

UNMANNED MISSIONS

(A Reference Book Based on Press Releases of NASA)

RATNESH DWIVEDI

*To Pragun*

## Chapters

**Acknowledgement**

**Forward**

**CHAPTER -1 : History of JPL**

**CHAPTER-2: Spacecraft in Heliocentric Orbit**

**CHAPTER-3: History of Comets and Asteroids**

**CHAPTER -4: Recent Historical Missions of NASA**

**HAPTER-5: My Note About Red Planet and NASA's Various Missions to Mars**

**CHAPTER -6: Cassini Huyegence Mission and Probe to Explore Ringed Planet(Saturn)**

**CHAPTER-7: Colombia:STS-107**

## Acknowledgement

In this untiring work to study seven selected mile stones of NASA,author is thankful to NASA at large,President Bush and his administration, American Astronomical Society and NASA Press Releases which made this work possible and which author has stored in his mail box since the day STS -107 collapsed during its re-entry in Earth's atmosphere in 2003.

Lastly he is thankful to his parents,his siblings,his wife and his little son who is now old enough to undersatnd that his father is mad about space sciences.

# Foreword

Space Sciences is future of mankind not because they explore space beyond planet earth but because they have answer to un-unswered and unknwons of universe.Who knew that Lord of Rings had more than 70 moons prior to 1997-2017(The Duration of Cassinii Hyugence Mission and Probe),Who knew that Mars has icy surface before Spirit and Opportunity and who knew that NASA will embard to a new space age with launch test of Orion Spaceship.From testing light weight missile launchers to space craft- space shuttle to space ship NASA has seen a long history of up and downs through space war age.In 60 years of its establishment it has set up many mile stones and has produces many noble brains,has answered many queries of small kids to mature scientists to noble laurets.

My passion is space science started when I was seven an my mother would narrate stories from Hindu Mythology and about Sun,Star and Moons.This contiinued iin my subconciious only to be triggered in 2003 when I saw STS-107 collapsing during its reenetry to Earth killing all seven astronauts on board including India born Kalpana Chawla. The one thing which I did next day as sitting on computer and signing the e-maiil updates which I contiinue till date now in 2017.It produced exceptional amount of curiosity in me to know more and that led attending NASA workshops and organizing NASA Scientist for a Day Contest between 2012-2-17 in Indian schools.

In this compiled work I have selected few recent mile stones of NASA and want to present before you so you also can know that these seven selected milestones are going to shpae the world for next 200 years. Trust me or not but its true.

History of world's most famous laboratory Jet Propulsion Laboratory will keep inspiring mankiind for next millenium.All facts and technology of a Spae Craft is tested in Helio centric orbit,comets and asteroids will keep increasing curiosity and challenges of man kind for all time.The only mission of NASA to Sun,The genesis,though faiiled by few hundred feets away from earth but had already done a remarkable job. Missions to Mars and Saturn are two most successful missions of NASA and have opened a new world for scientists.STS-107 has changed the ay we thiink about space shuttles and is a start point for a new space age.

Lastly these few milestones selected here for study are going keep mankind hooks for centuaris to come.

Wish you all the best at each mile stone.

Ratnesh Dwivedi

Noida ,India

September 21,2017

# CHAPTER -1 : History of JPL

NASA Facts

National Aeronautics and

Space Administration

Jet Propulsion Laboratory

California Institute of Technology

Pasadena, CA 91109

**Jet Propulsion Laboratory**

A new generation of space missions to explore the solar system and the universe beyond is unfolding at the Jet Propulsion Laboratory .

The American space age began January 31, 1958, with the launch of the first U.S. satellite, Explorer 1, built and controlled by JPL. In the four decades since then, JPL has led the world in exploring all of the solar system's known planets, except Pluto, with robotic spacecraft. The tools developed at JPL for its spacecraft expeditions to other planets have also proved invaluable in providing new insights and discoveries in studies of Earth, its atmosphere, climate, oceans, geology and the biosphere.Entering the new millennium as the 21st century begins, JPL continues as a world leader in science and technology, breaking new ground in the miniaturization and efficiency of spacecraft components. At the same time, the Laboratory is pushing the sensitivity of space sensors and broadening their applications for a myriad of scientific, medical, industrial and commercial uses on Earth.JPL is a federally funded research and development facility managed by the California Institute of Technology for the National Aeronautics and Space Administration..

JPL's Beginnings

JPL's history dates to the 1930s, when Caltech professor Theodore von Karman oversaw pioneering work in rocket propulsion. Von Karman was head of Caltech's Guggenheim Aeronautical Laboratory. Several of his graduate students and assistants gathered to test a primitive rocket engine in a dry riverbed wilderness area in the Arroyo Seco , a drycanyon wash north of the Rose Bowl in Pasadena, California. Their first rocket firing took place there on October 31, 1936.

After the Caltech group's successful rocket experiments, von Karman, who also served as a scientific adviser to the U.S. Army Air Corps, persuaded the Army to fund development of strap-on rockets (called "jet-assisted take-off') to help overloaded Army air-planes to take off from short runways. The Army helped Caltech acquire land in the Arroyo Seco for test pits and temporary workshops. Airplane tests at nearby air bases proved the concept and tested the designs. by this time, World War **II** had begun and the rockets were in demand.

As the group wound up the work on the jet-assisted takeoff rockets, the Army Air Corps asked von Karman for a technical analysis of the German V-2 program just discovered by Allied intelligence. He and his research team then proposed a U.S. research project to understand, duplicate and reach beyond the guided missiles beginning to bombard England. In the proposal, the Caltech team referred to their organization for the first time as "the Jet \Propulsion Laboratory."

Funded by Army Ordnance, the Jet Propulsion Laboratory's early efforts would eventually involve technologies beyond those of aerodynamics and propellant chemistry, technologies that would evolve into tools for space flight, secure communications, spacecraft navigation and control, and planetary exploration.

The team of about 100 rocket engineers began to expand, and the team began testing in the California desert of small unguided missiles (named Private) that reached a range of nearly 18 kilometers (about 11 miles). They experimented with radio telemetry from missiles, and began planning for ground radar and radio sets. By 1945, with a staff approaching 300, the group had begun to launch test vehicles from White Sands, New Mexico, to an altitude of 60 kilometers (200,000 feet), monitoring performance by radio.

Control of the guided missile was the next step, requiring two-way radio as well as radar and a primi- tive computer (using radio tubes) at the ground sta- tion. The result was JPL's answer to the German V-2 missile, named Corporal, first launched in May 1947, about two years after the end of war with Germany.

Developing a missile that would fly and survive in the field involved testing its aerodynamic design and durability under vibration and other stresses. The

team developed a supersonic wind tunnel and an array of environmental test technologies, all of which had wider use and came to support outside customers. Developing so complex a device as a missile to fly unaided and beyond reach of repair meant a new degree of quality, new test techniques and a new discipline called system engineering.

Subsequent Army work further sharpened the technologies of communications and control, of design and test and performance analysis. This made it possible for JPL to develop the flight and ground systems and finally to fly the first successful U.S. space Mission, Explorer 1. The entire three-month effort began in November 1957 and culminated with the successful launch on January 31, 1958.

On December 3, 1958, two months after NASA was created by Congress, JPL was transferred from Army jurisdiction to that of the new civilian space agency. It brought to the new agency experience in building and flying spacecraft, an extensive background in solid and liquid rocket propulsion systems, guidance, control, systems integration, broad testing capability, and expertise in telecommunications using low-power spacecraft transmitters and very sensitive Earth-based antennas and receivers.

The Laboratory now covers some 72 hectares (177 acres) adjacent to the site of von

Karman's early rocket experiments. Jet propulsion is no longer the focus of JPL's work, but the world-renowned name remains the same.

## Planetary **Exploration**

In the 1960s, JPL began to conceive and execute robotic spacecraft to explore other worlds. This effort began with the Ranger and Surveyor missions to the Moon, paving the way for NASA's Apollo astronaut lunar landings. During that same period and through the early **1970s,** JPL carried out Mariner missions to Mercury, Venus and Mars.

Mariner 2 became the first spacecraft to fly by another planet following its launch August 27, 1962, to Venus (Mariner 1 was lost because of a launch vehicle error). Other successful Mariners included Mariner 4, launched in 1964 to Mars; Mariner 5, launched in 1967 to Venus; Mariner 6, launched in 1969 to Mars; Mariner 7, launched in 1969 to Mars; and Mariner 9, launched in 1971 to orbit Mars.

Mariner 10 became the first spacecraft to use a "gravity-assist" boost from one planet to send it on to another — a key innovation in spaceflight that would later enable the exploration of outer planets that would have otherwise been unreachable. Mariner 10's launch in November 1973 delivered the spacecraft to Venus in February 1974, where a gravity-assist swing by allowed it to fly by Mercury in March and September that year .The first search for life on Mars was conducted in 1975 when NASA launched the Viking mission's two orbiter spacecraft and two Martian landers. The elaborate mission was divided between several NASA centers and private U.S. aerospace firms, with JPL building the Viking orbiters, conducting mission communications and eventually assuming responsibility for management of the mission.

Credit for the single mission that has visited the most planets goes to JPL's Voyager project.

Launched in 1977, the twin Voyager 1 and Voyager 2 spacecraft flew by the planets Jupiter (1979) and Saturn (1980-81). Voyager 2 then went on to an encounter with the planet Uranus in 1986 and a flyby of Neptune in 1989. Early in 1990, Voyager 1 turned its camera around to capture a series of images assembled into a "family portrait" of the solar system. Still communicating their findings as they speed out toward interstellar space, the Voyagers are expected to communicate information about the Sun's energy field until perhaps the second decade of the 21st century. In February 1998, Voyager 1 passed NASA's Pioneer 10 to become the most distant human-made object in space.In 1989 and 1990 NASA's Space Shuttle helped launch three JPL-managed solar system exploration missions: Magellan to Venus, Galileo to Jupiter and Ulysses to study the Sun's poles.

Magellan used a sophisticated imaging radar to pierce the cloud cover enshrouding Venus and map the planet's surface. Magellan was carried into Earth orbit in May 1989 by Space Shuttle Atlantis. Released from the shuttle's cargo bay, Magellan was propelled by a booster engine toward Venus, where it arrived in August 1990. It completed its third 243-day period mapping the planet in September 1992. Magellan mapped variations in Venus's gravity field before the mission ended in October 1994. At the conclusion of the mission, flight controllers corn- manded Magellan to dip into the atmosphere of Venus in a test of aerobraking — a technique for using atmospheric drag to slow spacecraft that has since been used in other planetary missions.The Galileo mission began October 18, 1989, when it was carried into Earth orbit on Space Shuttle Atlantis and then sent on its interplanetary flight path via ail, Inertial Upper Stage booster. Relying on gravity-assist swingbys to reach Jupiter, Galileo flew past Venus once and Earth twice. Along the way Galileo flew by the asteroid Gaspra in October 1991 and the asteroid Ida on August 28, 1993. On its final approach to the giant planet, Galileo observed Jupiter being bombarded by fragments of the broken-up comet Shoemaker-Levy 9. On July 12, 1995, Galileo separated from its atmospheric probe and the two spacecraft flew in formation to their final destination.

On December 7, 1995, Galileo fired its main engine to enter Jupiter orbit and collected data radioed from the probe during its parachute descent into the planet's atmosphere. During its two-year prime mission, Galileo conducted 10 targeted flybys of Jupiter's major moons. In December 1997, the spacecraft began a first extended mission, which featured eight flybys of Jupiter's icy moon Europa and two of the volcanic moon Io. A second extended mission began in early 2000, including flybys of the moons Io, Ganymede and Callisto, plus coordinated observations with the Cassini spacecraft as Cassini flew past Jupiter in December 2000 for a gravity assist to reach Saturn. Galileo's final flyby, of the small moon Amalthea, left the orbiter on course for an intentional impact into Jupiter in September 2003.N

ASA's shuttle fleet again launched a probe bound for other parts of the solar system when the Space Shuttle Discovery carried aloft Ulysses in October 1990. A joint mission between NASA and the European Space Agency, this project for the first time sent a spacecraft out of the ecliptic — the plane in which Earth and other planets orbit the Sun — to study the Sun's north and south poles. Ulysses first flew by Jupiter in February 1992, where the giant planet's gravity flung it into an unusual solar orbit nearly perpendicular to the ecliptic plane.

The prime mission concluded in September 1995. Ulysses continued to monitor the Sun as the spacecraft passed again over the south pole in September 2000 and over the north pole a year later.

The mission of Mars Observer, launched aboard a Titan III rocket September 25, 1992, ended with disappointment in August 1993 when contact was lost with the spacecraft shortly before it was to enter orbit around Mars. Some science instruments from Mars Observer are currently being reflown on Mars Global Surveyor.The next JPL planetary launches were those of Mars Global Surveyor and Mars Pathfinder, launched in November and December 1996, respectively. Mars Pathfinder put a lander and rover on the surface of the red planet in a highly successful landing July 4, 1997; the project fulfilled all the objectives of its prime mission and lasted considerably longer than originally designed before the lander fell silent in September 1997. Mars Global Surveyor went into orbit around the red planet on September 12, 1997 (September 11 EDT/PDT), and spent a year and a half lowering its orbit using the technique of aerobraking. The spacecraft began its prime mission in spring 1999 and is currently making highly detailed maps of the Martian surface.A disappointment at Mars occurred in late 1999, however, with the loss of the orbiter and lander developed and launched under the Mars '98 project — named Mars Climate Orbiter and Mars Polar Lander, respectively. Climate Orbiter entered the planet's atmosphere too low and did not survive orbit insertion on September 23, 1999. Polar Lander and two Deep Space 2 microprobes piggybacking on it to Mars were lost during arrival at the planet December 3, 1999.A orbiter named Mars

Odyssey was launched on April 7, 2001, and arrived at the red planet on October 24 of that year (October 23 EDT/PDT). The spacecraft embarked on a four-year examination of what Mars is made of, detecting water and shallow buried ice, and studying the radiation environment in space. The latest Mars mission is a pair of rovers launched in June and July 2003. With far greater mobility than 1997's Mars Pathfinder rover, these robotic explorers will be able to trek up to 40 meters (about 44 yards) across the surface each Martian day. Later proposed missions in the Mars program include

a powerful reconnaissance orbiter in 2005, small projects called "Scout" missions beginning in 2007, and a surface mission called the Mars Science Laboratory in 2009, leading eventually to a sample return mission.

JPL designed and built the Cassini mission to Saturn, launched on October 15, 1997. Cassini is carrying a probe, Huygens, provided by the European Space Agency, which will descend to the surface of Titan, Saturn's largest moon, six months after Cassini's July 2004 arrival at the ringed planet. Titan appears to boast organic chemistry possibly like that which led to the existence of life on Earth. Cassini flew by Venus in April 1998 and June 1999, followed by an Earth flyby in August 1999 and a Jupiter flyby in December 2000.

In 1995, NASA selected a JPL-teamed proposal to develop and fly a mission called Stardust under the space agency's Discovery program of low-cost missions. Launched in February 1999, Stardust will fly past comet Wild-2 in January 2004 and collect dust and volatile materials. Those materials will be returned to Earth in a capsule that will parachute to a landing on a dry lake bed in Utah in 2006.JPL is also providing project management for another Discovery mission, Genesis. Launched in August 2001, Genesis is collecting samples of charged particles in the solar wind and will return them to Earth laboratories in September 2004 for detailed analysis. In addition, JPL is managing a Discovery mission called Deep Impact that will propel a large copper projectile into the surface of comet Tempel 1, creating a huge crater expected to reveal information about the composition and structure of the comet nucleus. Deep Impact is planned for launch in 2004 with comet arrival in mid-2005. A comet is also the destination for a JPL instrument called the Microwave Instrument on the Rosetta Orbiter, or Miro, which will be carried by a European Space Agency craft due for launch in 2003.

Two asteroids will be the destination for Dawn, the first spacecraft ever planned to orbit two different bodies after leaving Earth. This mission under NASA's Discovery program will launch in 2006, orbit Vesta in 2010 and 2011, then begin orbiting Ceres in 2014.

Another major initiative for a new breed of NASA spacecraft is New Millennium, designed to flight-test new technologies so that they can be reliably used in science missions of the 21st century.

The first New Millennium spacecraft, Deep Space 1, was launched in October 1998 to test an ion engine and 11 other new technologies. Deep Space 1 tested autonomous navigation and two advanced science instruments when it flew by the asteroid 9969 Braille on July 29, 1999 (July 28 PDT). After its primary mission, Deep Space 1 gathered images and other information from a bonus September 22, 2001, flyby of comet Borrelly. Under Deep Space 2, two microprobes to test the Martian soil for water vapor piggybacked on Mars Polar Lander, but wore lost at arrival in December 1999. The'New Millennium program also includes deep space and Earth orbiting missions managed by other NASA centers.In addition to directing spacecraft that visit planets, asteroids and comets, JPL scientists are active in many programs of observations from the ground. The Laboratory created the Near-Earth Asteroid Tracking system, an automated system used at an Air Force observatory in Hawaii to scan the skies for asteroids or comets that could threaten Earth. In 1999, the project made its first observations from a second site, the 1.2-meter-diameter (48-

inch) Oschin telescope on Palomar Mountain, California. In 1998, NASA designated JPL as home of the agency's Near-Earth Objects Office to coordinate observations of Earth-crossing asteroids and comets by various NASA scientists.

**Earth Science**

In the late 1970s, JPL engineers and scientists realized that the sensors they were developing for interplanetary missions could be turned upon Earth itself to better understand our home planet. This has led to a series of highly successful Earth-monitoring missions that have evolved into a major segment of the Laboratory's activities, now sponsored by NASA's Office of Earth Science.

In 1978, JPL built an experimental satellite called Seasat to test a variety of oceanographic sensors including imaging radar, altimeters, radiometers and scatter meters. Many of the later Earth-monitoring instruments developed at JPL owe their legacy to the Seasat mission.

The imaging radar flown on Seasat led to a pair of missions flown on the Space Shuttle, 1981's

Shuttle Imaging Radar-A and 1984's Shuttle Imaging Radar-B. These were followed by Spaceborne Imaging Radar-C, an experiment teamed with the Gerrna,n/Italian X-Band Synthetic Aperture Radar and flown on the Space Shuttle twice in 1994. This mission's goal was to study a variety of scientific disciplines — geology, hydrology, ecology and oceanography — by comparing the radar images to data collected by teams of people on the ground. Imaging radar was reflown on the Space Shuttle under the Shuttle Radar Topography Mission in February 2000 to create the world's most accurate topographic map .Sea sat also tested an altimeter that measured sea level heights from space. This concept led to a full-scale satellite mission developed jointly by JPL and the French space agency, Topex/Poseidon. The ocean graphic satellite, launched August 10, 1992, on an Ariane 4 rocket from Kourou, French Guiana, has provided scientists with unprecedented insight into global climate and ocean interactions, currents, eddies, and new details about the global ocean seafloor. Jason 1, a follow-on mission to Topex/Poseidon and another U.S.-French collaboration, was launched December 7, 2001, and is currently in orbit.

Another mission with heritage in Seasat is the JPL-built NASA Scatterometer, an instrument that measures near-surface ocean winds from space. This instrument was launched in August 1996 on the Advanced Earth Observing Satellite prepared by Japan's National Space Development Agency, and continued operating until the satellite failed in early 1997. JPL prepared a rapid replacement, QuikScat, which was launched in 1999 from Vandenberg Air Force Base, California, carrying a scatterometer called SeaWinds. A follow-on SeaWinds scatterometer was launched on a Japanese satellite in 2002.

JPL also designed and built an instrument called the Microwave Limb Sounder that studies the chemistry of Earth's upper atmosphere, relaying important data on topics such as ozone depletion. Early versions flew as payloads on the Space Shuttle, followed by an instrument onboard NASA's Upper Atmosphere Research Satellite launched in September 1991. Currently, a new-generation version of the instrument is being developed to fly on a satellite for launch in 2004 on NASA's Aura satellite as part of the agency's Earth Observing System program.

JPL is responsible for several other instruments being flown under the Earth Observing System program. They include the Multi-angle Imaging Spectro Radiometer, launched on NASA's Terra satellite in December 1999 to study the role of clouds in global climate; the Advanced Spaceborne Thermal Emission and Reflection Radiometer, also carried on the Terra satellite to image Earth in various parts of the color spectrum; the Atmospheric Infrared Sounder, launched aboard NASA'S Aqua satellite in May 2002 , which relays data on temperature and humidity in the atmosphere in order to help understand how heat is exchanged between land, air, sea and the atmosphere; and the Tropospheric Emission Spectrometer, planned for launch in 2004 aboard the Aura satellite, which will help scientists understand the causes of acid rain and track trends in atmospheric chemistry on a global scale.The Active Cavity Radiometer Irradiance Monitor, or Acrim, is an instrument that measures the Sun's total output of optical energy from ultraviolet to infrared wavelengths -- called the total solar irradiance -- an important factor in the study of Earth's climate. The instrument was flown on several shuttle missions and satellites in the 1980s and 1990s. A dedicated satellite called Acrim Sat was launched in December 1999.Clouds will be the object of study for a trio of satellites called Cloud Sat planned for launch in 2004. They will be the first spacecraft to examine clouds on a global basis. A JPL-teamed mission called the Gravity Recovery and Climate Experiment, or Grace, launched twin satellites in March 2002 to conduct global high-resolution studies of Earth's gravity field. JPL provided project management for the Solar Mesosphere Explorer, a satellite launched in 1981 to investigate the processes that create and destroy ozone in Earth's upper atmosphere.

**Astrophysics**

In addition to studying Earth itself and other bodies within the solar system, other JPL missions extend our view deeper into the universe. JPL developed and manages the Space Infrared Telescope Facility, a sibling to the Hubble Space Telescope. The Space Infrared Telescope Facility, along with Hubble, the Chandra X-ray Observatory and the Compton Gamma-ray Observatory, are all part of NASA's Great Observatories Program, designed to study the universe at various wavelengths. The Space Infrared Telescope Facility, scheduled for launch in August 2003, will use infrared technology to study celestial objects that are either too cool, too dust-enshrouded or too far away to otherwise be seen. The observatory will pierce the thick dust that permeates the universe to unveil new information about galaxies, stars and dusty discs around nearby stars, which may be "planetary construction zones."The Space Infrared Telescope Facility builds on JPL's rich history of involvement with infrared astronomy. JPL in conjunction with Caltech operates the Infrared Processing and Analysis Center, which has processed data for several infrared missions, including the Infrared Astronomical Satellite and the ground-based Two Micron All-Sky Survey. JPL was U.S. manager of the Infrared Astronomical Satellite, a joint project with the Netherlands and the United Kingdom. Launched in 1983, the mission was an Earth-orbiting telescope that mapped the sky in infrared wavelengths. Its data have led to a wealth of discoveries about the formation of galaxies, stars and planets, including the first-ever direct evidence of an emerging planetary system around a star besides the Sun - material orbiting Vega, 26 light-years away. The Two Micron All-Sky Survey was astronomy's most thorough high-resolution digital survey of the entire sky, completed by twin infrared telescopes. Operations began in Arizona in June 1997 and in Chile in March 1998, and observations concluded in February 2001. The survey produced catalogues brimming with nearly half a billion objects. The bonanza of astronomical discoveries includes hun scientists to define new classes of stars; maps of the Milky Way's structure and dust distribution, and large-scale structure in the nearby universe; and observations of numerous dust-obscured galaxies in the distant universe.Following the Space Infrared Telescope Facility mission, JPL will provide crucial instrumentation for three upcoming infrared missions. JPL is partnering with the European Space Agency to develop the Mid-Infrared Instrument for the James Webb Space Telescope, a successor to Hubble that will look back in time over more than 90 percent of the history of the universe. JPL is also providing instrumentation to two missions of the European Space Agency, the Herschel Observatory and the Planck project to study the cosmic background. In 1996, NASA assigned JPL programmatic responsibility for the space agency's-Origins program. The program ties together a variety of spacecraft missions and instruments to study the formation of galaxies, stars, planets, and life. Origins seeks to answer the fundamental questions: Where did we come from? Are we alone?The Origins Program at JPL includes the currently operating Keck Interferometer, a system developed by JPL that links two 10-meter (33-foot) telescopes atop Mauna Kea, Hawaii, to form the world's most powerful ground-based telescope system. The system combines data from the telescopes to create the equivalent of a single, larger telescope. It is capable of looking for large planets in dust clouds around nearby stars, and will pave the way for future Origins missions. The Space Interferometry Mission, being devel- oped for launch in 2009, will link multiple telescopes in space to search for planets around other stars, from the size of Jupiter down to planets a few times as massive as Earth. In addition, the mission will carry out a program of fundamental astrophysics by mea- suring the distances and motions of tens of thousands of individual objects in our own and nearby galaxies. JPL manages the Kepler mission on behalf of a science team based at NASA's Ames Research Center. Kepler, with a planned 2007 launch, is designed to look for planets that cross in front of, or "transit," other stars. Kepler may find evidence for hundreds of Earth-sized planets. The premier mission in the Origins program is the proposed Terrestrial Planet Finder which, following a launch around 2015, would look directly for Earth-like planets in the habitable zones of nearby stars. Terrestrial Planet Finder would analyze the chemistry of the...atmospheres of these terrestrial planets to look for signs of life.In April 2003, NASA launched the Galaxy Evolution Explorer. This Caltech-JPL mission, developed under NASA's Small Explorer Program, uses ultraviolet wavelengths to study the history of star formation. It will observe a million galaxies across 10 billion years of cosmic history to help astronomers determine when the stars we see today had their origins. JPL designed and built the Wide Field/Planetary Camera, the main observing instrument on NASA's Hubble Space Telescope. After a flaw was discovered in the space telescope's main mirror, JPL created a second-generation camera, the Wide Field and Planetary Camera 2, that compensated for the optical problem - essentially like fitting Hubble with a set of corrective eyeglasses. Spacewalking astronauts installed this second camera during a shuttle mission in December 1993, allowing Hubble to fulfill its promise in producing unprecedented views of the cosmos.

**Telecommunications**

To provide tracking and communications for planetary spacecraft, JPL designed, built and operates NASA's Deep Space Network of antenna stations. Communications complexes are located in California's Mojave Desert, in Spain and in Australia. In addition to NASA missions, the network regularly performs tracking for international missions sending spacecraft to deep space. Ground stations also conduct experiments using radar to image planets and asteroids, as well as experiments using the technique of very long baseline interferometry to study extremely distant celestial objects. The Deep Space Network is also playing a major role in Space Very Long Baseline Inter ferometry, a radio astronomy project teaming orbiting spacecraft


with ground antennas. Combining ground antennas with a Japanese spacecraft launched in

1997 approximately triples the resolving power previously achievable.

**Technologies**

In the three decades it has led the nation's planetary exploration program, JPL has honed several skills and areas of innovation, including deep space navigation and communication, digital image processing, imaging systems, intelligent automated systems, instrument technology, microelectronics and more. Many of these disciplines found applications outside the planetary spacecraft field, from solar energy to medical imagery.In the mid-1970s, in'response to a world energy crisis, JPL worked to develop and apply alternate sources of electricity such as solar energy, for the Department of Energy, and electric vehicles and other alternative transport systems, for the Department of Transportation.The Laboratory has also applied space-based operational, communication, and information processing techniques to the needs of the Department of Defense, Federal Aviation Administration and other federal agencies. Its active technology transfer program with the industrial community dates back to the early days of the missile program.JPL conducts technology development projects both for NASA and for sponsors other than NASA. Non-NASA projects at JPL have included Firefly, an aircraft-borne infrared fire mapping system for the U.S. Forest Service; a document monitoring system to help the National Archives safeguard the U.S. Constitution, Declaration of Independence and Bill of Rights; medical projects such as robot-assisted microsurgery and medical imaging systems, and Internet-based telemedical systems; and varied projects in such fields as advanced spacecraft and sensor technology, microelectronics, supercomputing and environmental protection. JPL work for the Department of Defense has included the Miniature Seeker Technology Integration, a satellite built and launched in November 1992 to demonstrate miniature sensor technology and a rapid development system. JPL also managed the U.S. Army's All Source Analysis System project, a battlefield information management system.

Research and development activities at JPL include an active program of automation and robotics supporting planetary rover missions and NASA's Space Station program. In supercomputing, JPL has pioneered work with new types of massively parallel computers to support processing of enormous quantities of data to be returned by space missions in years to come.

**Institutional**

In addition to the Laboratory's main Pasadena site and the three Deep Space Network complexes around the world, JPL installations include an astronomical observatory at Table Mountain, California, and a launch operations site at Cape Canaveral, Florida.

In 2003, JPL has a workforce of about 5,500 employees and on-site contractors, and an annual budget of approximately $1.4 billion.

Dr. Charles Elachi, a scientist with a background in imaging radar and other remote-sensing technologies, became director of JPL on May 1, 2001. In addition to his JPL post he serves as a vice president of Caltech. Elachi's predecessors as head of the Laboratory were Dr. Edward C. Stone (1991-2001), **Dr.** Lew Allen Jr. (1982-1990), Dr. Bruce **Murray (1976-1982),** Dr. William H. Pickering (1954-1976), Dr. Louis Dunn (1946-54), Dr. Frank Malina (19441946) and Dr. Theodore von Karman (1944 and forerunner organization).

## **JPL Spacecraft Missions**

*Spacecraft, Launch Date, Mission Description, Comment* Explorer 1, 1/31/58, first U.S. satellite, operated to 5/23/58 Explorer 2, 3/5/58, satellite, launch failed

Explorer 3, 3/26/58, satellite, operated to 6/16/58 Explorer 4, 7/26/58, satellite, operated to 10/6/58 Explorer 5, 8/24/58, satellite, launch failed

Pioneer 3, 12/6/58, escape attempt, in orbit to 12/7/58

Pioneer 4, 3/3/59, escaped to solar orbit, tracked to 650,000 km (400,000 mi)

Ranger I, 8/23/61, lunar prototype, launch failure Ranger 2, 11/18/61, lunar prototype, launch failure

Ranger 3, 1/26/62, lunar probe, spacecraft failed, missed Moon

Ranger 4, 4/23/62, lunar probe, spacecraft failed, impact

Ranger 5, 10/18/62, lunar probe, spacecraft failed, missed

Ranger 6, 1/30/64, lunar probe, impact, cameras failed

Ranger 7, 7/28/64, lunar probe, successful, 4,308 pictures

Ranger 8, 2/17/65, lunar probe, successful, 7,317 pictures

Ranger 9, 3/21/65, lunar probe, successful, 5,814 pictures

Surveyor 1, 5/30/66, lunar lander, operated 6/2/66-1/7/67

Surveyor 2, 9/20/66, lunar lander, crashed 9/23

Surveyor 3, 4/17/67, lunar lander, operated 4/20-5/4/67

Surveyor 4, 7/14/67, lunar lander, crashed 7/17

Surveyor 5, 9/8/67, lunar lander, operated 9/11-12/17/67

Surveyor 6, 11/7/67, lunar lander, operated 11/10-12/14/67

Surveyor 7, 1/7/68, lunar lander, operated I/10-2(21/68

Mariner 1, 7/22/62, Venus probe, launch failed

Mariner 2, 8/27/62, Venus flyby 12/14/62, signal lost 1/3/63

Mariner 3, 11/5/64, Mars probe, shroud failed

Mariner 4, 11/28/64, Mars flyby 7/14/65 with pictures, signal lost 12/20/67

Mariner 5, 6/14/67, Venus flyby 10/19/67

Mariner 6, 2/24/69, Mars flyby 7/31/69 with pictures, lasted to 12170

Mariner 7, 3/27/69, Mars flyby 8/5/69 with pictures, lasted to 12/70

Mariner 8, 5/8/71, failed Mars launch

Mariner 9, 5/30/71, Mars orbiter 11/13/71 to 10/27/72

Mariner 10, 11/3/73, Venus swingby 2/5/74, Mercury 3/29, 9/21, 3/16/75

Viking 1, 8/20/75, Mars orbiter/lander, orbit 6/19/76, landing 7/20/76

Viking 2, 9/9/75, Mars orbiter/lander, orbit 8/7/76, landing 9/3/76

Voyager 1, 9/5/77, Jupiter 3/5/79, Saturn 11/12/80 with pictures, continuing on interstellar mission

Voyager 2, 8/20/77, Jupiter 7/9/79, Saturn 8/25/81, Uranus 1/24/86, Neptune 8/25/89, continuing on interstellar mission

Seasat, 6/27/78, ocean radar satellite, operated three months Solar Mesosphere Explorer, 10/6/81, successful

Infrared Astronomical Satellite, 1/25/83, NASA/United Kingdom/Netherlands orbiting infrared telescope, operated to 11/23/83

Magellan, 5/4/89, Venus radar mapper, orbited 8/10/90 - 10/13/94, mapped 99% of planet

Galileo, 10/18/89, Jupiter orbiter/probe; Venus swingby 2/10/90, Earth swingby 12/8/90, asteroid Gaspra flyby 10/29/91, second Earth swingby 12/8/92, Ida flyby 8/28/93, Shoemaker-Levy observations 7/94; arrived at Jupiter 12/7/95 for two year mission and accomplished atmospheric probe portion of mission; extended mission focused on Jupiter's moons Europa and Io; Jupiter impact planned 9/21/03

Ulysses, 10/6/90, European Space Agency/NASA solar polar mission; Jupiter swingby 2/8/92, solar southern polar passages 1994 and 2000, northern polar passages 1995 and 2001

Mars Observer, 10/25/92, lost at Mars orbit insertion (8/24/93)

Topex/Poseidon, 8/10/92, U.S.-French ocean satellite, operating

Mars Global Surveyor, 11/7/96, entered Martian orbit 9/12/97, science mission began 3/99, operating

Mars Pathfinder, 12/4/96, landed 7/4/97 and deployed instrumented rover

Cassini, 10/15/97, Saturn orbiter with Huygens descent probe to study Saturn's moon Titan; Venus flybys 4/26/98 and 6/24/99, Earth flyby 8/18/99, Jupiter flyby 12/30/00, Saturn arrival 7/1/04, Huygens descent 1/14/05

Deep Space 1, 10/24/98, tested ion engine and 11 other advanced technologies; asteroid flyby 7/99, comet flyby 9/01

Mars Climate Orbiter, 12/11/98, lost during Mars arrival 9/23/99

Mars Polar Lander and Deep Space 2 microprobes, 1/3/99, lost during Mars arrival 12/3/99

Stardust, 2/7/99, en route to comet flyby 1/2/04, Earth return 1/15/06

Wide-field Infrared Explorer, 3/4/99, telescope coolant lost shortly after launch

Quick Scatterometer (QuikScat), 6/19/99, ocean winds satellite, operating

Active Cavity Irradiance Monitor Satellite (AcrimSat), 12/20/99, Earth-orbiting satellite monitoring Sun's radiation output, operating

2001 Mars Odyssey, 4/7/01, entered orbit 10/24/01, science mission began 2/02, operating

Genesis, 8/8/01, solar wind sample return, Earth return in 9/04 Jason 1, 12/7/01, U.S.-French ocean satellite, operating

Gravity Recovery and Climate Experiment, 3/17/02, twin satellites mapping Earth's gravity, operating

Mars Exploration Rover 'Spirit,' 6/10/03, instrumented rover, en route to landing 1/4/04

Mars Exploration Rover 'Opportunity,' 7/7/03, instrumented rover, en route to landing 1/25/04

## **Major Instruments on Other Spacecraft**

*Instrument; Launch Date; Host Spacecraft; Purpose/Comment*

Wide Field/Planetary Camera, 4/24/90, NASA's Hubble Space Telescope, main camera for orbiting telescope

Wide Field and Planetary Camera 2, 12/2/93, NASA's Hubble Space Telescope, camera that solved telescope's optical problem

NASA Scatterometer, 8/17/96, Japan's Advanced Earth Observing Satellite, mapped sea winds through early 1997

Multi-angle Imaging Spectro-Radiometer, 12/18/99, NASA's Terra satellite, Earth imaging

Advanced Spaceborne Thermal Emission and Reflection Radiometer, 12/18/99, NASA's Terra satellite, Japan-U.S. land-mapping instrument

Atmospheric Infrared Sounder, 5/4/02, NASA's Aqua satellite, weather and climate studies

SeaWinds, 12/13/02, Japan's Midori satellite, ocean winds instrument, operating

## **JPL Payloads on Space Shuttle Missions**

Payload Name; Shuttle Flight; Launch/Landing Dates; Purpose

Shuttle Multispectral Infrared Radiometer; STS-2; November 12-14, 1981; remote identification of rock-forming minerals

Shuttle Imaging Radar-A; STS-2; November 12-14, 1981; imaging radar

Fiber Optic Data Transmission Experiment; STS-41C; April 6-13, 1984; measure long-term radiation effects on communication systems in experiment on Long Duration Exposure Facility deployed on this mis¬sion; retrieved on mission STS-32, January 9-20, 1990

Shuttle Imaging Radar-B; STS-41G; October 5-12, 1984; imaging radar

Atmospheric Trace Molecule Spectroscopy Experiment; STS-51B, April

29-May 6, 1985; STS-45, March 24-April, 1992; STS-56, April 8-17,

1993; STS-66, November 3-14, 1994; measurement of trace and minor

gases in Earth's atmosphere

Dexterous End Effector Experiment; STS-62; March 4-18, 1994; to test sensor on shuttle robotic arm

Spacebome Imaging Radar-C/X-13and Synthetic Aperture Radar; STS-59, April 9-20, 1994; STS-68, September 30-October 11, 1994; imaging radar

Cryo System Experiment; STS-63; February 3-14, 1995; test cryocooler and heat pipe

KidSat; STS-76, March 22-31, 1996; STS-81, January 12-22, 1997; STS-86, September 25-October 6, 1997; sharing Earth images online with schools

Inflatable Antenna Experiment; STS-77; May 19-29, 1996; test antenna structure designed to be inflated in orbit

Confined Helium Experiment; STS-87; November 19-December 5, 1997; high-resolution test of the theory of confined systems

Electronic Nose (E-nose); STS-95; October 29-November 7, 1998; air quality technology inspired by human nose architecture

Shuttle Radar Topography Mission; STS-99; February 11-22, 2000; high-resolution database of Earth topography

# CHAPTER-2: Spacecraft in Heliocentric Orbit

The *following table lists spacecraft in heliocentric* othit soiled *by spacecraft name,* **SO FAR FRM SPECIFIC SPACE ORGANISATIONS-**
*Spacecraft* **Name Sponsor Launch Date Mission Description**

*Giotto* ,E-3A 1985-07-02 Halley
**SOHO** ESA 2'995-12-02 Solar observatory at Sun-
**Earth LI point**
***Nozomi (Planet-131** Japan 1998-07-04 Mars orbit*
*Sak.gahe        Japan 1985-01-07 Halley*
*&sisal  Japan 1985-08-18 Halley*
*ACE       US 1997-08-25 Spgce weather observatory at Sun-Ea;117* **LI point**
**Cassini**        US **1997-1045** *Saturn orbit*
**Clementine** *.;  US 1994-01-25 Moir Geogrephes*
**Deep Space**   *US 1998-10.24 Asteroid flyby*
*GalileoUS 1989-10-18 Jupiter* orbit
*Helios 1       US 1974-1240 Solar orbit*
*Helios 2       US 1976-0145 Solar orbit*
*ISEE" 3 ./ ICE US 1978-0842 Sun-Earth LI point,,*
*Giacobini-2.'hyne,*

***Magellan*** *US **989-05-08** Venus omit*
***Mariner -10*** *1973-11-03 VLs., Mercury*
*Mariner 2* 7 Venus
*Mariner 4* *1964-11-28* Mars
*Mariner 5* *967-06-14* 'Omits
*Mariner 6* *1969-02·24 Mars*
*Mariner 7* *1969-03-27* Mars
*Mariner 9* *US 1971-05-30 Mars orbit*
US 1998-12-11 Mars orbit

*Man; Climate Orbiter*
 19904147 Mars orerff

  1992-09-25 Mars orbit
  199E-12-04 Mars Ian zAt
Mars Global   US
*Surveyor*
***Mars** Observer **Mars** Pathfinder*

|  |  |
|---|---|
| ;W,755PolarLat*N?ep | VS 199tL0 1*413* Mars *landing* **ptobe** **US 1996-02-17 Eros** *orbit* |
| Space2 | US 1972-03-03 Jupiter, SS **esca** |
| elonpef | :94-06 Jupiter, **Saturn,** |
| Pioneer 7-f | **Lis 1959-03-03 Missed moon** |
| Pioneer 4 | US **1960-03-11 Solar** *orbit* |
| Pioneer 5 | US 1965-12-16 Solar orbit. |
| Pioneer 5 | US **1966-08-17 Solar orbit** |
| Pioneer 7 | !PS **1967-12-13 Solar orbit** |
| Pioneer 8 | *l* 1968 11 08 Solar orbit |
| Pioneer 9 | |
| Pioneer Vein's | ice., tts landin |

|  |  |  |
|---|---|---|
| | US | 196,, |
| obe Bus Ranger | vs- | 1962-1 |
| | | Cooret flyby |
| 1.13 | 1990-11,; | |
| ifs | 1975 08-20 | 4. |
| US | | Mars orbit, |
| US | 1975-00-09 | |
| US | 12F 7-09 05MEWS *orbit*, lailefi |
| | 1977-08-20. | |
| US | l 959-01-02 | Jupiter, Saturn, SS |
| USSR | 1971-05-19 | jupiterlig.?,penne, |
| USSR | 1971-05-28 | |
| USSR | 1973-0741 | 11Thg.d owa l |
| e!ISSR | 1973-07-25 | ix10·, Ampace |
| USSR | 1973-0V-08 | Mars orbit, land |
| USSR | 1173.06 09 | |

*3* Ranger 5 Stardust

Ulysses Viking 1 Viking 2 Voteagtiy Vi7yager 2
Luna 1 Mars 2 Mars 3 Mars 4 Mars 5 Mars 6 Mars 7 Phobos 1 Phobos 2
ega I Vega 2
energy 10 Vedera 11 Vaneri-·,· sera 13 Veneit 14 Venera 15 Venera 16 Venera 1 VitWeira **VenerR** 3 tie/let:a 4


*USSR 1988-070-⁷ 114,,an3 wtit*
*USSR 1988-07-1.2 Afros orbit*
*USSR 198442-15 Venus landing, Halley*
**USSR** *1984-12-21 Venus landing, Halley USSR97⁻ ,,·7,-06....14 Venus orbit,. landing USSR 1976-00- 09 Venus: landing 1155R 1978-09-14 Venus landing USSR 1981-10-$0 Venus landing*
*USSR 1981-11-V4 Venus landing USSR 1983-06- 02 Venus orbit USSR 1983-06-07 Venus orbit*
**USSR 1961-62-12** *Versus*
**USSR** *1965-1 1- 12 tie,nue.;*
*USSR 1965-11·16 Venus        t*
*USSR I 96 7 -06; 1,2 t4Nius' impact*

*Earth: Geodetic and Geophysical Vara*

| PHYSICAL PARAMETER | VALUE |
|---|---|
| ass | 5.9730 × 10²⁴ kg |
| 'Wean raditik, v | 6371 a o,,o2 krn |
| Density | 5,515 g cm^-3 |

*ATOMAL RAIVILIS:,*

*4 Y9761*     *6376140 m*

*(Geod. ref. sys., 1980)*
*(Merit, 1983)*

***FLATTENING**, fa-bya (IA141 976; Merit)*

*(G. AV, S,,1980)*

***GRAVITY.' 2 COEF,***

*(IAU,1 976)*

*(GEM 72,1911 C0*

*c.22 522 (B-A)/41 a A2*

*Longitude ';f axis a*

*SURFACE GRAVITY 9 P*

• *P*
•

*5111/1? VA2*

*637813
7   177
637813
6 m*

*1./2982
57*

*1/29825
722'
6356.75
2 kro*

*0.00108
263*

*0.00108
26265
1,5744x
10A-6 -
0,9038x
10A-6*

*7.26
15x10'6*

14.9285 deg

9.8321863685 m sA-2
9.7803267715 sA-2
9.82022 m sA-2

Preoesshm constant

H J2
IW a'20
3,,
27376341
x10A-3

C (Polar moment)
0.3307007,41

*0.3295181 Ati A2*

*0.3286108 Ma*

*Moan monyent I 0.3299765 M a $^{r4}$2*

*03307144 M RoA2*

**Mean rotation rate-**$_9$ **omega** *7,292 t 15x10 A-5 rayt's*

*r ega$^{ii}$2 1/$^4$3/61K,aarth*

*a A3/G,M earth 11288,901*

*tatic J 2h 0,0010722*

*.G 73 tief h 1299.66*

**Fluid** *core* **ratifus(PREM)**
*3480 km*

*Inner core* **radius**
*1215 km*

*¹¹·⁷A8S OF LA MRS A .6170Spheirl?*

n    eans

**Crust**

**Mantle**

**Outer core Inner core**

*5..1 )10¹'18 kg 1,4k10.⁻'21kAY*

*.2-.16x10⁻⁴22 kg $043x1(?' $4 kg 183510A24 kg 9.675..*10⁴.22 kg*

## MOMENTS OF INERTIA

| | |
|---|---|
| Mantle, $I_m/M_{earth} a^2$ | 0.29234 |
| Fluid core, $I_f/M_{earth} a^2$ | 0.03760 |
| Hydrostatic, $(C_f - A_t)/C_f$ | 1/393.10 |
| Observed, $(C_f - A_f)/C_f$ | 1/373.81 |
| Hydrostatic, $(C_{ic} - A_{Ic})/C_{ic}$ | 1/416 |

*Free core nutation period*          $429.8\ d$

*Chandler wobble period*            $434.3\ d$

*SURFACE AREA*

| | |
|---|---|
| *Land* | $1.48 \times 10^8\ km^2$ |
| *Sea* | $3.62 \times 10^8\ km^2$ |
| *Total* | $5.10 \times 10^8\ km^2$ |

# CHAPTER-3: HISTORY OF COMETS AND ASTEROIDS

Solar System Exploration TarcretBodies
Here is a list of asteroid and comet targets   recent and future solar system exploration missions. The list is sorts:: Oy encounter date.
Target  Body  Encounter  Mission  Launch Date
Asteroid 252 Mathilde
*(body* data, *geocentric elpherr* 1997-Jun-27 f

AsterOID 9969 BRAILLE

(b0dy data geocentric ephemeris 1999-Jul-28  21hy  *PSI*  1998-Oct-25 asteroid 433 Eros

*(body data, oeocentric ephemeris)* *2000-Feb-14 rendezvous NEAR 1996 Feb-17*
Comet 19P/E-. - ly

(hoc *data,* ocentric *ephemeris 2001-Sep-22 flyby DS1 1998-Oct-25*
*Annefrank*
(**body** data, tric Fi7memel, s 203 _-?Jot'-3 fly*by Stardust 1999-Fet-*
06
*Comet S1P/W*
'body data, 02
*2006-Jmn-* 5˜tar(‚ust L.)·), 1$^1$-06
*Comet 9P/Temoel*
(body *data, aeocentri˜ 2005-Jul-04 impact flyby*
*Deep Impact*
*2004-Dec-$0*
*Asteroid 25143 1998 SEM)*
*(body data, aeocentric ephemeris) 2005-Jun-15 arrival*
*2005-Nov-02 departur*
*2007-jun-10 Earth-return MUSES-C 200-May-09*
Asteroid *2867 Steins*
*(body* data, geocentric *ephemeri,, 4) 2008-Sep-05 flyby Rosetta 2004-Mar02*
*Asteroid* 21 Lutetia -
*(body data, aeocentric ephemeris) 2010-Jul-10 flyby Rosetta 2004-Mar02*
*Comet* 67P/Churyumov-Geralmen
(body *data,* ge::centric *enj7emer* 2OI4 -Aug rendezvous: iosetta
*2004-17-02*
Credits

Contact: *Webmaster (webma-   (.jp1.nasa.ao¯v)*
*Last modified: 2004 May 28 11:37*

## GREAT COMETS IN HIISTORY

Donald K. Yeomans

Jet Propulsion Laboratory/California Institute of Technology

A relative few comets are so visually impressive as to be termed "great comets". Just the right set of circumstances must occur. Far from the sun, the solid portions of comets, which consist mostly of water ice and embedded dust particles, are inactive. They are not large enough to be seen with the naked eye. However, when near the sun, the icy cometary surfaces vaporize and throw off large quantities of gas and dust thus forming the enormous atmosphere and tails that make comets so visually striking. It is the fluorescing of these gases, and particularly the reflection of sunlight from the minute dust particles in the comet's atmosphere and tail, that can make these objects so visually impressive. However, this activity by itself does not insure that a comet will become a great comet. An active comet can only become great by making a particularly close approach to the sun so that it produces enormous quantities of gas and dust or by making a close approach to the Earth so that its tail can be easily viewed. In either case, great comets must be seen in a dark sky.

While applying the appellation "great comet" to a particular cometary return is a subjective process, the following Table is an attempt to list the great naked-eye comets that have been reported. With the single exception of periodic comet Halley, all the tabulated comets have passed through the inner solar system either for the first time or the intervals between their returns are measured in thousands or millions of years.

The first tabular entry gives the approximate date when the comet was first reported as a naked-eye object. The following column gives the approximate observational interval (in days) during which the comet remained a naked eye object. The next two columns give the date and distance in astronomical units when the comet reached its closest point to the sun (perihelion). One astronomical unit is the mean distance between the sun and Earth. The following columns give the date and distance when the comet reached its closest point to the Earth (perigee), and the date and apparent magnitude when the comet reached its brightest in a dark sky. A diffuse cometary image becomes noticeable to the naked eye when it reaches a magnitude of approximately 3.4 in a dark sky. Compared to a comet whose magnitude is 4, a 3rd magnitude comet would appear 2.5 times brighter and a magnitude 2 comet would appear 2.5 x 2.5 = 6.3 times brighter still, etc. The brightest star in the sky (Sirius) has an apparent magnitude of -1.5. At its brightest, the planet Jupiter appears at magnitude -2.7.

**1st Date          Obs         Perihelion                    Perigee**

Reported Int Date Dist Date Dist Date Mag Name, Notes

YYYY/mmm/DD (d) YYYY/mmm/DD (AU) YYYY/mmm/DD (AU) YYYY/mmm/DD

*Julian Calendar*

## B.0 DATES

**373-372 Winter**

87/Jul 35        87/Aug/06 0.59              87/Jul/27 0.44 87/Ju
1P/Halley

12/Aug/25     57    12/Oct/10 0.59        12/Sep/10 0.16
1P/Halley

A.D DATES

66/Jan/30        71    66/Jan/26 0.59        66/Mar/20 0.25
1P/Halley

141/Mar/26 41 141/Mar/22 0.58 141/Apr/22 0.17 141/Apr/22 -1 1P/Halley

989/Aug/10 32    989/Sep/05              0.58
    989/Aug/20 0.39   989/Aug/20 1-2

1P/Halley

1066/Apr/02 66    1066/Mar/20            0.58
    1066/Apr/24 0.10   1066/Apr/24 -1

1P/Halley

1106/Feb/02 40             4

1132/Oct/03 24    1132/Aug/30            0.74
    1132/Oct/07 0.04   1132/Oct/07 -1

(1132 T1)

1145/Apr/15 65    1145/Apr/18            0.58
    1145/May/12 0.27   1145/May/12

0    1P/Halley

1222/Sep/02 36    1222/Sep/28            0.58
    1222/Sep/06 0.31   1222/Sep/24

1-2 1P/Halley, 5

1240/Jan/27 64    1240/Jan/21            0.67
    1240/Feb/02 0.36   1240/Feb/02 0

(1240 B1)

1264/Jul/17

(1264 N1), 6 85    1264/Jul/20           0.82
    1264/Jul/29 0.18 1264/Jul/29 0

1301/Sep/01 61    1301/Oct/25           0.58
    1301/Sep/23 0.18    1301/Sep/23

1-2 1P/Halley

1378/Sep/26 15    1378/Nov/10           0.58
    1378/Oct/03 0.12    1378/Oct/03 1

1P/Halley, 7

1402/Feb/08 70    1402/Mar/21           0.38
    1402/Feb/19 0.71    1402/Mar/12 -

3    (1402 D1),    8

1456/May/26 44    1456/Jun/09           0.58
    1456/Jun/19 0.45    1456/Jun/19 0

1P/Halley

1468/Sep/18 56    1468/Oct/07           0.85
    1468/Oct/02 0.67    1468/Oct/02 1-

2 (1468 Si)

1471/Dec/25 59    1472/Mar/01         0.49
    1472/Jan/23 0.07    1472/Jan/23 -3

(1471 Y1)

1531/Aug/05 34    1531/Aug/26         0.58
    1531/Aug/14 0.44    1531/Aug/27

1    1P/Halley

1532/Sep/02 120    1532/Oct/18        0.52
    1532/Sep/21 0.67    1532/Oct/13-

1    (1532 R1)

1533/Jun/27 83    1533/Jun/15         0.25
    1533/Aug/02 0.42    1533/Jun/27 0

(1533 M1), 9

1556/Feb/27 72    1556/Apr/22         0.49
    1556/Mar/13 0.08    1556/Mar/14 -2

(1556 D1)

1577/Nov/01 87    1577/Oct/27         0.18
    1577/Nov/10 0.63    1577/Nov/08 -

3    (1577 V1)

178/Sep    80

191/Oct           2

218/May    40    218/May/17         0.58
    218/May/30 0.42    218/May/30 0

1P/Halley

240/Nov/10 39    240/Nov/10         0.37
    240/Nov/30 1.00    240/Nov/20  1-

2 (240 V1)

295/May

1P/Halley    30    295/Apr/20         0.58
    295/May/12 0.32    295/May/12 0

374/Mar/03 32    374/Feb/16         0.58
    374/Apr/02 0.09    374/Apr/02 -1

1P/Halley

390/Aug/21 26    390/Sep/05         0.92
    390/Aug/18 0.10    390/Aug/18 -1

(390 Q1), 2

400/Mar/18 30      400/Feb/25                0.21
         400/Mar/31 0.08     400/Mar/19 0

(400 Fl)

442/Nov/09 2 (442 V1)

451/Jun/09 1P/Halley

565/Jul/22    100

68

100    442/Dec/15    1.53    451/Jun/28    0.58
565/Ju1/15 0.82     442/Dec/07              0.58
451/Jun/30 0.49 565/Sep/13 0.54 442/Dec/07 1-
451/Jun/30 0 565/Sep/13 0-1

(565 01)

568/Jul/28    106    568/Aug/27              0.87
         568/Sep/25 0.09     568/Sep/25 0

(568 01)

607/Mar-Apr 30     607/Mar/15               0.58
         607/Apr/19 0.09     607/Apr/19 -2

1P/Halley

684/Sep/06  33     684/Oct/02                0.58
      684/Sep/07 0.26    684/Sep/07 1-2

1P/Halley

760/May/16  50     760/May/20                0.58
      760/Jun/03 0.41    760/Jun/03 0

1P/Halley

770/May/25  62     770/Jun/05                0.58
      770/Jul/I0 0.30    770/Jul/10 1-2

(770 K1)

837/Mar/21  39     837/Feb/28                0.58
      837/Apr/11 0.03    837/Apr/11 -3

1P/Halley, 3

838/Nov/09  49

891/May/12  62           2

905/May/18  26     905/Apr/26                0.20
      905/May/25 0.21    905/May/23 02

Gregorian Calendar

1618/Nov/16 67    1618/Nov/08              0.40
      1618/Dec/06 0.36    1618/Nov/29

0-1 (1618 W1)

1664/Nov/17 75    1664/Dec/04              1.03
      1664/Dec/29 0.17    1664/Dec/29

1      (1664 W1)

1665/Mar/27 24    1665/Apr/24              0.11
      1665/Apr/04 0.57    1665/Apr/20 -1

(1665 F1), 10

1668/Mar/03 27    1668/Feb/28              0.07
      1668/Mar/05 0.80    1668/Mar/08

1-2 (1668 El)

1680/Nov/23 88    1680/Dec/18              0.01
      1680/Nov/30 0.42    1680/Dec/29

1-2 (1680 V1),    11

1682/Aug/15 41    1682/Sep/15              0.58
      1682/Aug/31 0.42    1682/Aug/31

0-1 1P/Halley

1686/Aug/12 34     1686/Sep/16            0.34
     1686/Aug/16 0.32    1686/Aug/27

1-2 (1686 R1)

1743/Nov/29 110    1744/Mar/01            0.22
     1744/Feb/27 0.83    1744/Feb/20 -

3     (1743 X1), 12

1769/Aug/24 94     1769/Oct/08            0.12
     1769/Sep/10 0.32    1769/Sep/22 0

Messier (1769 P1), 13

1807/Sep/09 90     1807/Sep/19            0.65
     1807/Sep/27 1.15    1807/Sep/20


1-2 Great Comet (1807 R1)

1811/Apr/11     260     1811/Sep/12     1.04
1811/Oct/16 1.22 1811/Oct/20 0 Great Comet (1811 F1)

1843/Feb/05     48     1843/Feb/27     0.006
1843/Mar/06  0.84  1843/Mar/07  <-3  Great March Comet (1843 D1), 14

1858/Aug/20 80 1858/Sep/30 0.58 1858/Oct/11 0.54 1858/Oct/07 0-1 Donati (1858 L1)

1861/May/13 90 1861/Jun/12 0.82 1861/Jun/30 0.13 1861/Jun/27 0 Great Comet (1861 J1), 13

1865/Jan/17 36 1865/Jan/14 0.03 1865/Jan/16 0.94 1865/Jan/24 1 Great Southern Comet (1865 B1), 15

1874/Jun/10 50 1874/3ul/09 0.68 1874/Jul/23 0.29 1874/Jul/13.`0-1 Coggia (1874 H1)

1882/Sep/01 135 1882/Sep/17 0.008 1882/Sep/16 0.99 1882/Sep/08 <-3 Great September Comet (1882 R1), 16

**1901/Apr/12 38 1901/Apr/24 0.24 1901/Apr/30 0.83 1901/May/05 1 Great Comet (1901 Gl)**

**1910/Jan/13 20 1910/Jan/17 0.13 1910/Jan/18 0.86 1910/Jan/30 12 Great January Comet (1910 Al), 17**

**1910/Apr/10 80 1910/Apr/20 0.59 1910/May/20 0.15 1910/May/20 0-1 1P/Halley**

1927/Nov/27 32 1927/Dec/18 0.18 1927/Dec/12 0.75 1927/Dec/08 1 Skjellerup-Maristany (1927 X1), 18

1965/Oct/03 30 1965/Oct/21 0.008 1965/Oct/17 0.91 1965/Oct/14 2 Ikeya-Seki (1965 S1), 19

1970/Feb/10 80 1970/Mar/20 0.54 1970/Mar/26 0.69 1970/Mar/20 0-1 Bennett (1969 Y1), 20

1976/Feb/05 55 1976/Feb/25 0.20 1976/Feb/29 0.79 1976/Mar/01 -1 West (1975 V1), 21

1996/Mar/15 30 1996/May/01 0.23 1996/Mar/25 0.10 1996/Apr/20 1-2 Hyakutake (1996 B2)

1996/Sep/09 215 1997/Apr/01 0.91 1997/Mar/22 1.32 1997/Mar/26 -0.7 Hale-Bopp (1995 01), 22

---

Notes

Reported by the Greek historian Ephorus to have split into two pieces. The Chinese reported that the tail spanned more than 70 degrees.

The closest approach to the Earth that comet Halley has ever made. On Apr. 13, the comet's tail was more than 90 degrees in length.

This comet passed very close to the sun and is perhaps the progenitor of the sungrazing comets of 1882 and 1965 or that of 1843.

Korean observers reported the comet was visible during the daylight hours on September 9th (probably during twilight only).

On July 26, Chinese observers reported the tail spanning 100 degrees.

Chinese observers reported cloudy weather from October 11 until Nov. 9, at which time the comet had passed behind the sun.

In mid-March, the comet entered solar conjunction and there were reports that it was a daylight object for 8 days.

The comet was discovered emerging from solar conjunction. Last observed on April 20 as it approached solar conjunction.

This was the first comet discovered with the aid of a telescope (on Nov. 14). Visible in daylight only 12 degrees from the Sun on February 27.

Tail reported as longer than 90 degrees near Earth close approach.

On the date of perihelion, this sungrazing comet was observed in daylight nearly one degree from the sun.

Comet observed in southern hemisphere.

The Great September comet was a brilliant object that was observed very close to the sun, and split into at least four separate pieces near perihelion. This comet and comet Ikeya-Seki in 1965 are believed to be members of the same family of sungrazing comets.

This comet was easily observed on January 17 only 4.5 degrees from the sun. It is often confused with the later apparition of comet Halley in mid-1910.

On December 18, this comet was seen in daylight only 5 degrees from the sun. At the end of December, the tail was reported to be nearly 40 degrees in length.

Sungrazing comet Ikeya-Seki split into two or possibly three pieces near perihelion. Toward the end of October, the impressive tail reached lengths in excess of 45 degrees.

The tail of comet Bennett reached 10 degrees in mid-March.

Comet West's impressive broad tail reached a length of 30 degrees on March 8. Near perihelion, the comet split into four pieces.

The observational interval is based on the time during which the comet had a total magnitude of 3.4 or brighter.

# CHAPTER -4: *RECENT HISTORICAL. MISSIONS OF NATIONAL AREONAUTICS AND SPACE ADMINSTRATION*

**GENESIS MISSION TO EXPLORE SUN**

Savage (202) 358-1727

NASA Headquarters, Washington, D.C.

In a dramatic ending that marks a beginning in scientific research, NASA's Genesis spacecraft is set to swing by Earth and jettison a sample return capsule filled with particles of the Sun that may ultimately tell us more about the genesis of our solar system.

"The Genesis mission -- to capture a piece of the Sun and return it to Earth -- is truly in the NASA spirit a bold, inspiring mission that makes a fundamental contribution to scientific knowledge," said Steven Brody, NASA's program executive for the Genesis mission, NASA Headquarters, Washington.

On September 8, 2004, the drama will unfold over the skies of central Utah when the spacecraft's sample return capsule will be snagged in midair by helicopter. The rendezvous will occur at the Air Force's Utah Test and Training Range, southwest of Salt Lake City.

"What a prize Genesis will be," said Genesis Principal Investigator Dr. Don Burnett of the California Institute of Technology, Pasadena, Calif. "Our spacecraft has logged almost 27 months far beyond the moon's orbit, collecting atoms from the Sun. With it, we should be able to say what the Sun is composed of, at a level of precision for planetary science purposes that has never been seen before."

The prizes Burnett and company are waiting for are hexagonal wafers of pure silicon, gold, sapphire, diamond and other materials that have served as a celestial prison for their samples of solar wind partides. These wafers have weathered 26-plus months in deep space and are now safely stowed in the return capsule. If the capsule were to descend all the way to the ground, some might fracture or break away from their mountings; hence, the midair retrieval by helicopter, with crew members including some who have performed helicopter stunt work for Hollywood.

"These guys fly in some of Hollywood's biggest movies," said Don Sweetnam, Genesis project manager at NASA's Jet Propulsion Laboratory in Pasadena, Calif. "But this time, the Genesis capsule will be the star." The Genesis capsule -- carrying the agency's first sample return since the final Apollo lunar mission in 1972, and the first material collected beyond the Moon -- will enter Earth's atmosphere at 9:55 am Mountain Time.

Two minutes and seven seconds after atmospheric entry, while still flying supersonically, the capsule will deploy a drogue parachute at 33 kilometers (108,000 feet) altitude. Six minutes after that, the main parachute, a parafoil, will deploy 6.1 kilometers (20,000 feet) up. Waiting below will be two helicopters and their flight crews looking for their chance to grab a piece of the Sun.

"Each helicopter will carry a crew of three," said Roy Haggard, chief executive officer of Vertigo Inc. and director of flight operations for the lead helicopter. "The lead helicopter will deploy an eighteen-and-a-half foot long pole with what you could best describe as an oversized, Space-Age fishing hook on its end. When we make the approach we want the helicopter skids to be about eight feet above the top of the parafoil. If for some reason the capture is not successful, the second helicopter is 1,000 feet behind us and setting up for its approach We estimate we will have five opportunities to achieve capture."

The helicopter that does achieve capture will carry the sample canister to a dean room at the Michael Army Air Field at the U.S. Army Dugway Proving Ground, where scientists await their cosmic prize. The samples will then be moved to a special laboratory at NASA's Johnson Space Center, Houston, where they will be preserved and studied by scientists for many years to come.

"I understand much of the interest is in how we retrieve Genesis," added Burnett. "But to me the excitement really begins when scientists from around the world get hold of those samples for their research. That will be something."

JPL, a division of the California Institute of Technology, manages the Genesis mission for NASA's Science Mission Directorate, Washington. Lockheed Martin Space Systems, Denver, developed and operates the

spacecraft. Los Alamos National Laboratory and NASA's Johnson Space Center contributed to Genesis payload development, and the Johnson Space Center will curate the sample and support analysis and sample allocation.

More detailed background on the mission is available at http://genesismission.jpl.nasa.gov .

---

DC Agle (818) 393-9011

Jet Propulsion Laboratory, Pasadena, Calif

Donald Savage (202) 358-1547

NASA Headquarters, Washington, D.C.

RELEASE: 2004-217 September 7, 2004

**NASA a 'Go' For Midair Capture of Samples from the Sun**

NASA's Genesis spacecraft crossed the orbit of the Moon early Monday, Sept. 6, on its way to the mission's dramatic finale over the skies of west-central Utah tomorrow. Genesis, bringing back samples of the solar wind, is NASA's first sample return mission since Apollo 17 returned the last of America's lunar samples to Earth in December 1972.

An important milestone in the mission was met Monday morning, when the Genesis spacecraft performed its final trajectory maneuver before capsule release and the dramatic midair capture over Utah. The spacecraft passed the Earth-Moon orbit at about 2 a.m. Pacific Time on Monday, traveling at about 1.25 kilometers per second (2,700 miles per hour).

"Our Deep Space Network is allowing us to keep a close eye on our spacecraft and its samples of the Sun," said Genesis project manager Don Sweetnam of NASA's Jet Propulsion Laboratory in Pasadena, Calif. "It is right where we planned it to be. Everything is go. The navigators and engineers here at JPL are go, and the recovery team out in Utah is go, too."

The Genesis recovery team members, both ground support and the flight crews who will make the dramatic midair capture, have been undergoing flight training since arriving at the U.S. Army Dugway Proving Grounds, Utah, on Aug. 23.

"We came here with a specific set of mission goals that had to be met before Sept. 8, and those have all been met or exceeded," said Genesis director of flight operations Roy Haggard of Vertigo Inc., Lake Elsinore, Calif. "The next time these two helicopters take to the sky one of them will be landing with a spacecraft hooked to its belly."

The Genesis sample return capsule will enter Earth's atmosphere at 8:55 a.m. Pacific Time over Oregon. Two minutes and one time zone later, the capsule will deploy its drogue parachute at 33 kilometers (108,000 feet) over the vast alkali flats and sagebrush of the U.S. Air Force's Utah Test and Training Range. Waiting 29.5 kilometers (97,000 feet) below will be two helicopters and crew bearing the space-age equivalent of a fisherman's rod-and-reel, ready to catch some Sun.

"From the time the drogue deploys it will take about 18 minutes for the capsule to reach a height where we can get to it," said Genesis prime pilot Cliff Fleming of South Coast Helicopters, Santa Ana, Calif. "When we are up there that may feel like a long 18 minutes but we have been training for this moment since 1999, so in the grand scheme of things another quarter-hour or so shouldn't matter much."

The Genesis mission was launched in August 2001 on a journey to capture samples from the storehouse of 99 percent of all the material in our solar system -- the Sun. The samples of solar wind particles, collected on ultra-pure wafers of gold, sapphire, silicon and diamond, will be returned for analysis by Earth-bound scientists. Thesamples Genesis provides will supply scientists with vital information on the composition of the Sun, and will shed light on the origins of our solar system.

samples Genesis provides will supply scientists with vital information on the composition of the Sun, and will shed light on the origins of our solar system.The Genesis events will be carried live on NASA Television Sept. 8 and will be webcast live at

http: / / www. j pl .nasa.goviwcbcasti.nesis/ .

JPL manages the Genesis mission for NASA's Science Mission Directorate, Washington. Lockheed Martin Space Systems, Denver, developed and operates the spacecraft. JPL is a division of the California Institute of Technology.

**MY NOTE ON GENESIS MISSION**
DEAR ALL,

The 215 MILLION DOLLAR GENESIS MISSION RE ENTERED AFTER A FOUR YEAR VOYAGE. TO STUDY INVISIBLE SOLAR WIND AND HELLUM REACTION INSIDE BURNING STAR THE MANY EXPERMENTS DURING ITS STAY BETWEEN EARTH AND SUN WERE, AWESOME AND SCIENTISTS AROUND THE WORLD STUDED SOME OF THE RARES RECTIONS INSIDE SUN SOLAR WIND WHICH

MAKE A VERY CLOSE AND PENETRATION EFFECTS ON EARTH FOR LONGER PERIODS WERE CAPTURED BY GENESIS CAPSUL.

CLOST TO THE DOORSTEP ALMOST A IT FIND ITSELEUNABLE TO OPEN ITS WING/PARACHUTES AND IMPACTED BY GENESIS CAPSUL.

CLOSE TO THE DOORSTEP ALMOST A IT FIND ITSELF UNABLE TO OPEN ITS WING/PARACHUTES AND IMPACTED OVER EARTH IN THE DESERT OF UTAH.

HERE U MAY FIND MORE STRAIGHT FROM NASA AND CLEAN ROOM.

RATNESH DWIVEDI

DC Agle (818) 393-9011

Jet Propulsion Laboratory, Pasadena, Calif.

Donald Savage (202) 358-1547

NASA Headquarters, Washington, D.C.

RELEASE: 2004-219 September 8, 2004

Genesis Mission

Status Report

The Genesis sample return capsule entered Earth's atmosphere at 9:52:47 a.m. Mountain Daylight Time and entered the preplanned entry ellipse in the Utah Test and Training Range as predicted. However, the Genesis capsule, as a result of its parachute not deploying, impacted the ground at a speed of 311 kilometers per hour (193 miles per hour). The impact occurred near Granite Peak on a remote portion of the range. No people or structures were anywhere near the area.

"We have the capsule," said Genesis project manager Don Sweetnam of NASA's Jet Propulsion Laboratory, Pasadena, Calif. "It is on the ground. We have previously written procedures and tools at our disposal for such an event. We are beginning capsule recovery operations at this time."

By the time the capsule entered Earth's atmosphere, the flight crews tasked to capture Genesis were already in the air. Once it was confirmed the capsule touched down out on the range, the flight crews were guided toward the site to initiate a previously developed contingency plan. They landed close to the capsule and, per the plan, began to document the capsule and the area.

"For the velocity of the impact, I thought there was surprisingly little damage," said Roy Haggard of Vertigo Inc., Lake Elsinore, Calif., who took part in the initial reconnaissance of the capsule. "I observed the capsule penetrated the soil about 50 percent of its diameter. The shell had been breached about three inches and I could see the science canister inside and that also appeared to have a small breach," he said.

The safety of recovery personnel has been the top priority. The capsule's separation charge had to be confirmed safe before the capsule could be moved. The recovery team is in the process of preparing to move the capsule to a clean room.

The Genesis mission was launched in August 2001 on a journey to capture samples from the storehouse of 99 percent of all the material in our solar system -- the Sun. The samples of solar wind particles, collected on ultra-pure wafers of gold, sapphire, silicon and diamond, were designed to be returned for analysis by Earth-bound scientists.

JPL manages the Genesis mission for NASA's Science Mission Directorate, Washington. Lockheed Martin Space Systems, Denver, developed and operated the spacecraft. JPL is a division of the California Institute of Technology.

For information about the Genesis Sample Return Mission on ;id-Internet, visit http://www.nasa.govigenesis . For background information about Genesis, visit http://genesisnussion.jpl.nasa.gov

---

September 8, 2004

Genesis Mission Status: Canister Transported

The science canister from the Genesis spacecraft has been transported by helicopter from its impact site at the U.S. Army Dugway Proving Ground in Utah to a holding area next to a specially constructed dean room on the Army base. The foil wrapping will be removed from the canister and dirt will be brushed off before the canister is moved into the clean room for analysis of the contents. NASA's Jet Propulsion Laboratory, Pasadena, Calif., manages the Genesis mission for NASA's Science Mission Directorate, Washington. Lockheed Martin Space Systems, Denver, developed and operated the spacecraft. JPL is a division of the California Institute of Technology.

For information about the Genesis Sample Return Mission on the Internet,

 visit http://www.nasa.gov/genesis

Large sections of the Genesis Sample Return Capsule that survived the hard impact landing in the Utah desert Sept. 8 were shipped to Lockheed Martin's facilities in Denver, Colo. and arrived Sept. 22. Members of the Genesis team and the Mishap Investigation Board have begun the process of closely inspecting and analyzing the capsule and the systems onboard to identify clues as to why the parachutes were not deployed from the capsule following re-entry into the earth's atmosphere. Meanwhile, members of the Genesis science team continue their meticulous work in a clean room at the

U.S. Army Dugway Proving Ground as they collect and secure the science canister and sample materials that were brought back in the capsule. Optimism continues to grow as the science team believes that a significant repository of solar wind materials survived the hard landing and has been recovered.

Credit: NASA/JSC

+ Medium Resolution

RELEASE: 2004-225 September 9, 2004

Genesis Mission Status: Conducting Inventory of Canister

Genesis team specialists are beginning the process of conducting inventory of the contents of the craft's science canister.

The canister is inside a Xclean room at the U.S. Army Dugway Proving Ground, Utah. Scientists are hopeful that the recovered Genesis samples will be sufficient to achieve the mission's science goals. The team is handling the canister and the sample return capsule in a methodical manner. When this inventory is completed, the materials will be transferred to NASA's Johnson Space Center, Houston, as originally planned.

NASA is convening a Mishap Investigation Board to determine why the Genesis drogue parachute failed to deploy.

---

DC Agie (818) 393-5011

Jet Propulsion Laboratory, Pasadena, Calif

Donald Savage (202) 358-1547

NASA Headquarters, Washington, D.C.

RELEASE: 2004-228  September 10, 2004

Genesis Scientists Bouncing Back From Hard Landing

Scientists who conducted the preliminary assessment of the Genesis canister are encouraged by what they see. They believe it may be possible t3 achieve the most important portions of their science objectives.

"We are bouncing back from a hard landing, and spirits are picking up again," said Orlando Figueroa, deputy associate administrator for programs for the Science Mission Directorate at NASA Headquarters in Washington.

"This may result in snatching victory from the jaws of defeat," added Dr. Roger Wiens of the Los Alamos National Laboratory in New Mexico, a member of the Genesis science team. "We are very encouraged." Based on initial inspection, it is possible a repository of solar wind materials may have survived that will keep the science community busy for some time.

"We are pleased and encouraged by the preliminary inspection," said NASA Administrator Sean O'Keefe. "The outstanding design and sturdy construction of Genesis may yield the important scientific results we hoped for from the mission."

"I want to emphasize the excellent work by the navigation team to bring the capsule back exactly on target was key in our ability to recover the science," said Andrew Dantzler, director of the Solar System Division at NASA Headquarters, Washington. "In addition, the robustness of the design of the spacecraft was the reason it could take such a hard landing and still give us a chance to recover the samples.

"The mission's main priority is to measure oxygen isotopes to determine which of several theories is correct regarding the role of oxygen in the formation of the solar system. Scientists hope to determine this with isotopes collected in the four target segments of the solar wind concentrator carried by the Genesis spacecraft. "From our initial look, we can see that two of the four concentrator segments are in place, and all four may be intact," Wiens said

.The mission's second priority is to analyze nitrogen isotopes that will help us understand how the atmospheres of the planets in our solar system evolved. "These isotopes will be analyzed using gold foil, which we have also found intact," Wiens said.

Other samples of solar winds are contained on hexagonal wafers. It appears these are all or nearly all broken, but sizable pieces will be recovered, and some are still mounted in their holders. "We won't really know how many can be recovered for some time, but we are far more hopeful important science can be conducted than we were on Wednesday," Wiens said.

Another type of collector material, foils contained on the canister's lid, were designed to collect other isotopes in the solar wind. It appears approximately three-fourths of these are recoverable, according to Dr. Dave Lindstrom, mission program scientist at NASA Headquarters. However, these foils have been exposed to elements of the Utah desert.

The Genesis mission was launched in August 2001 on a journey to capture samples from the storehouse of 99 percent of all the material in our solar system -- the Sun. The samples of solar wind particles, collected on ultra-pure wafers of gold, sapphire, silicon and diamond, were designed to be returned for analysis by Earth-bound scientists.

JPL manages the Genesis mission for NASA's Science Mission Directorate, Washington. Lockheed Martin Space Systems, Denver, developed and operated the spacecraft. JPL is a division of the California Institute of Technology.

For information about the Genesis Sample Return Mission on ;id-Internet, visit http://www.nasa.govigenesis . For background information about Genesis, visit http://genesisnussion.jpl.nasa.gov/ .

### Genesis **Mission Status: Canister Transported**

The science canister from the Genesis spacecraft has been transported by helicopter from its impact site at the U.S. Army Dugway Proving Ground in Utah to a holding area next to a specially constructed dean room on the Army base.

The foil wrapping will be removed from the canister and dirt will be brushed off before the canister is moved into the clean room for analysis of the contents.

NASA's Jet Propulsion Laboratory, Pasadena, Calif., manages the Genesis mission for NASA's Science Mission Directorate, Washington. Lockheed Martin Space Systems, Denver, developed and operated the spacecraft. JPL is a division of the California Institute of Technology.

For information about the Genesis Sample Return Mission on the Internet, visit ht.. :

The Genesis sample return capsule landed well within the projected ellipse path in the Utah Test & Training Range on Sept 8, but its parachutes did not open. It impacted the ground at nearly 320 kilometers per hour (nearly 200 miles per hour). NASA's Jet Propulsion Laboratory, a division of the California Institute of Technology in Pasadena, Calif., manages the Genesis mission for the agency's Science Mission Directorate. Lockheed Martin Space Systems, Denver, developed and operated the spacecraft.News and information about Genesis is available on the Internet at http ltwww.nasa.goylgenesis . For background information about Genesis, visit http://1yriesismission.jp1.nitsa.gov/ . For information about NASA on the Internet, visit http://www.nasa.gov/ .

---

NEWS RELEASE: 2004-231          September 16, 2004

Genesis Mission Status Report

Genesis team scientists and engineers continue their work on the mission's sample return canister in a specially constructed dean room at the U.S. Army Proving Ground in Dugway, Utah. As more of the capsule's contents are revealed, the team's level of "enthusiasm for the amount of science obtainable continues to rise. At present, the science canister that holds the majority of the mission's scientific samples is lying upside down -on its lid. Scientists are very methodically working their way "up" from the bottom portion of the canister by trimming away small portions of the canister's wall. The team continues to extract, from the interior of the science canister, small but potentially analyzable fragments of collector array material. One-half of a sapphire wafer was collected Tuesday - the biggest piece of collector array to date. The mission's main priority is to measure oxygen isotopes to determine which of several theories is correct regarding the role of oxygen in the formation of the solar system. Scientists hope to determine this with isotopes collected in the four target segments of the solar wind concentrator carried by the Genesis spacecraft. The condition of these segments will be better known over the next few days, after the canister's solar wind concentrator is extricated. At this time, it is believed that three of these segments are relatively intact and that the fourth may have sustained one or more fractures. There are no concrete plans regarding the shipping date of the Genesis capsule or its contents from Dugway to the Johnson Space Center in Houston. The team, continues its meticulous work and believes that a significant repository of solar wind materials may have survived that will keep the science community busy for some time. The Genesis sample return capsule landed well within the projected ellipse path in the Utah Test and

Training Range on Sept. 8, but its parachutes did not open. It impacted the ground at nearly 320 kilometers per hour (nearly 200 miles per hour).

For more information regarding the recovery and analysis of Genesis samples please contact Bill Jeffs of NASA Johnson Space Center at 281-483-5035 or via email at william. .ieffsfeinasti..p.v.

DC Agle (818) 393-9011

Jet Propulsion Laboratory, Pasadena, Calif.

Donald Savage (202) 358-1547

NASA Headquarters, Washington, D.C.

NEWS RELEASE: 2004-236September 23, 2004

Genesis Mission Status Report

The Genesis team has shipped its first scientific sample from the mission's specially constructed deanroom at the U.S. Army Proving Ground in Dugway, Utah. The sample, containing what are known as "lid foils," was attached to the interior lid of the Genesis sample return capsule.

This is the first batch in what we are growing more confident will be many more scientifically valuable samples," said Genesis Project Manager Don Sweetnam of NASA's Jet Propulsion Laboratory, Pasadena, Calif. "It appears that we have recovered about 75 to 80 percent of these lid foils. A great deal of credit has to go to the dedicated men and women of Genesis who continue to do very precise, detailed work out there in the Utah desert."

After the sample was shipped from Utah, it was received by Genesis co-investigator Nishizumi Kunihiko from the University of California, Berkeley, Space Sciences Laboratory.

In addition to the lid foils, there was optimistic news about the collector array. Team members from JPL arrived in Utah on Monday with a special fixture to aid in handling the science canister's stack of four collector arrays. The stack was successfully removed as one piece. With the stack on the fixture, the team has begun the process of disassembling the arrays. Several large pieces of individual collector materials, including one completely intact hexagon, were recovered from the top array.

# CHAPTER-5: MY NOTE ABOUT RED PLANET AND NASA'S VARIOUS MISSION TO MARS

THE 215 MILION DOLLAR GENESIS MISSION RE ENTERED AFTER A FOUR YEAR VOYAGE IN SOLAR WIND AND HELIUM REACTION INSIDE BURNING STAR.THE MANY EXPERMENTS DUE BETWEEN EARTH AND SUN WERE AWESOME AND SCIENTISTS AROUND THE WORLD STUDIED RARE RECTIONS INSIDE SUN. SOLAR WIND WHICH MAKE A VERY CLOSE AND PENETRATING EARTH FOR LONGER PERIOD WERE CAPTURED- BY GENESIS CAPSUL.CLOSE TO THE DOORSTEP ALMOST IT FIND ITSELF UNABLE TO OPEN ITS WING/PARTS OVER EARTH

IN THE DESERT OF UTAH. YOU MAY FIND MORE STRAIGHT NEWS FROM NASA AND CLEANROOM.

RATNESH DWIVEDI

SEPTEMBER 16/2004

TWO: A MISSION TO SECOND EARTH
   THE MISSION MARS

BEFORE YOU GO FOR DETAILED STUDY OF VARIOUS MISSIONS OF NASA TO EXPLORE RED PLANET I THINK ABOUT RED PLANET AND LIFE POSSIBILITIES OVER THERE.HINDUS HOLY EPIIC TALKS ABOUT ORIGIN ,EXPANSION OF UNIVERSE AND ALSO ABOUT SOLAR SYSTEM.

UNIVE.RSE HAS ALWAYES BEEN A MAJOR SOURCE OF ATTRACTION AND AMAZEMENT FOR ASTROPHYSICSTS,ASTROGEOLOGISTS AND COSMOLOGISTS.IT WAS NOT VERY LONG AGO I HAVING EXACT SCIENTIFIC DETERMINATION ABOUT ORIGIN AND EXPANSION OF UNIVERSE PHILOSOPHICAL DESCRIPTION WERE THERE/LIKE ONE WITH HINDU'S HOLY BOOK VEDA- YAJURVEDA .IT HAS DEFINITE HYMNS ABOUT UNIVERSE AND PRESENC OF LIFE ELEMENT ON RED PLANET EXPLAINING FEW HYMNS RELATED TO IT(HYMNS NOT GIVEN HERE DUE TO SPACE PROLEM.

IN HIS LARGER EXPANSION THE DIVINE BODY IS THE BEST POSSIBLE THING, EXTREMLEY DIFFICULT TO APPROACH. WITH THE HELP OF HIS MINUTEST ENERGY DIVINE BIRTH AND DISASTERS OF OTHER ENERGIES INSIDE IT.

THAT DIVANAL FORCE HAS CREATED THIS UNIVERSE AND MAKES ITS PRESENCE PRIOR TO WORLD. THE DIVINAL FORCE AT FIRST CREATED VAYU,THEN FIRE IT CREATD.THEN IT CREATED WATER AND IN THE WATER ITSELF IS MEDICINE AND TOGETHER WITH THEE LIFE PARTICLES LAST DIVINAL FORCE CREATED SOIL(WHERE FOOD GRAINS CAN BEGERMED.HE DIDIT IN VERY SYSTEMATIC MANNER WITHIN THIS UNIVERSE.THE UNIVERSE WHICH IS SATURATED AND FILLED WITH WATER AND THUNDER IN MIST OF DIVINAL ENERGY TO CREATE LIFE PARTICLES AND PROTECT THEM AT DISTANT

UNIVERSE THE DARK UNIVARSAL ENERGY WHICH PRIMARLY CONTAINS RED AND BLUE COLOR ALL FORMS IN LIVING CREATURES, ALWAYES TRY TO PROTECT US ON VARIOUS UNIVARSAL LAYERS. THE UNIVERSAL DIVINAL FOR CHANGEOWER THE DIVINAL WATER AT OUR DESIRED UNIVERE SHOW US A PATH OF HAPPINESS.

PRESENCE OF DIVINAL WATER SHOW US THE DIVINAL PATH AND SURLY BEARS THAT ALL AT DISTANT UNIVARSAL BODIES (POSSESING SAME CHARACTER AS OF EARTH)

MAY THE DIVINAL FORCE ACHIEVE US THE LIFE AND IMMORTALITY.LET US SEE HOW THE MYSTIQUE MARS HAS BEEN DESCRIBED.MERE THIS HYMN IS ENOUGH TO DESCRIBE LIFE PRESENCE ON PLANET MARS. IT HAS ALWAYES BEEN CMPARED A CLOSE CHARACTER TO EARTH. BE STUDIED

THAT MARS WAS PRODUCED WITH FRACTIONAL DIVISINOL OF EARTH.

AGNIRMOORDHA DIVHAKAKUTPATI PRITHIVYAAYAM,APAM RETA SI JINVATI.

MARS BEARS THUNDER AND WATER. IT DOES CNTAIN IN ITS YELLOWISH BROWNISH GRAM DERIVED FROM MOTHER EARTH. IT BEARS LIFE AND WISDOM AND A POWERFULL PLANE1 SIMILAR TO THAT OF EARTH.NOW SEE HOW BRILLIANTLY POSSIBILITIES OF ELEMENTRYLIFE HAS BEEN DESCRIBED WORSHIPING LORD SHIVA. WITH UNIVERSE,THE ELEMENTRY LIFE WHICH ORIGINATES IN DEEPER SPACE DARKER SPACE ON ENLIGHTEN EXISTS IN THUNDER,RAIN DROUGHT AND EVEN IN SNOW FALLS.LET US WORSHIP AND LABOUR FOR IT.IN DANGEROUS FALLS, IN DEEPER AND DENSEST FORESTS, IN NARROWY CANALS,FASTEST ELEMENTRY LIFE(RUDRA

SHAKTI)MAKES IT MIGHTY PRESENCE THRU DIVINAL ENERGY. LET US PRAY AND ADORE IT.

WITHIN THE FASTER FLOW OF UNIVERSAL ENERGY, IN THE TIME OF HAZARDS,HAVOCS, METALIC ELEMENTS DUMPED IN DEEPER SURFACES OF UNIVERSAL BODIES ,THIS ELMENTERY LIFE HAS MIGHTY PRESENCE.

THIS FRCTIONALISED AT WITH RAPID SPEED, IT MAY HAPPEN WITHIN DEEPER UNIVERSE . ENERGY NEVER TO BE LASTED.

LET US REALLY PRAY AND ADORE THE RUDRA SHAKTI THE SOURCE OF ELEMENTRY LIFE, MIGHTY PRESENCE ON LARGE UNIVARSAL DISTANCES AND BODIES.

RATNESH DWIVEDI

WINTERS OF 1999A.D

AYODHYA

RATNESH DWIVEDI SEPT 25, 2004

+91 080 252 87140

E-FAX---1-214-481-1501

Guy Webster (818) 354-6278

Jet Propulsion Laboratory, Pasadena, Calif.

Dwayne Brown (202) 358-1726

NASA Headquarters, Washington

News Release: 2004-161 June 25, 2004

Mars Rover Surprises Continue; Spirit, Too, Finds Hematite

On challenging slopes that NASA's Mars rovers began exploring this month,¡0 Opportunity have found new surprises for the folks back home. Spirit rolled up to a knobby rock just past where the "Columbia Hills" st surrounding plain. It touched the rock with a mineral-identifying instrugc arm and detected hematite. Hematite identified from orbit was NASA's keyi Opportunity's landing site halfway around Mars from these hills within Gua Opportunity, continuing its descent into "Endurance Crater," has found un between lower layers of rock it is examining for the first time and an ol§ Crater" where, months ago, the rover discovered evidence that water once "It's gratifying how well these machines keep performing, considering theb their original three-month missions on Mars," said Chris Voorhees, rover engineer at NASA's Jet Propulsion Laboratory, Pasadena, Calif. By the endL have worked on Mars for half a year. It has driven more than three times one kilometer (0.6 mile). The only symptom of wear or aging on either roll] friction in one wheel on Spirit. The rover team at JPL is beginning to cob solar-powered robots to spend the period of martian winter when reduced di power supply to a minimum. In the nearer term, though, team members are e through on the new scientific findings.

Spirit's hematite finding is in a rock dubbed "Pot of Gold," about the sif a shape as if somebody took a potato and stuck toothpicks in it, then putn the toothpicks," said Dr. Steve Squyres of Cornell University, Ithaca, Ng the rovers' science instruments. "How it got this crazy shape is anyone'sa a good theory yet."Dr. Doug Ming, a rover science-team member from NASA's Johnson Space Cent "There's apparently some type of weathering, a removal of material, but vg determine whether it's by chemical or mechanical processes."Further study of Pot of Gold could also help scientists assess what the .1⁻,u environmental conditions. "Hematite can form in a few different ways. Mos but it can also result from a dry, thermal oxidation process," Ming said.a from orbit that made Meridiani Planum a compelling place to send Opportun learned that the hematite is indeed part of a water story. At Gusev we're After examining Pot of Gold with the microscopic imager and two spectrome rover backed away from the rock to re-approach at a better angle for using expose the rock's interior. In the rough and slippery terrain, that manetr Other nearby rocks may also be inspected before Spirit resumes longer driv Columbia Hills area. Also, engineers are planning an attempt to redistri4 right front wheel before the rover leaves its current vicinity.

Team members presented both rovers' status at a press conference at JPL d driven far enough into the stadium-sized Endurance Crater to put it within layers of rock beneath a sulfate-rich layer. That area is similar to whatp the shallower "Eagle Crater," where it landed in January. "We're trying t the stratigraphy of the crater as we drive down, analyzing each unit cheri6 with all the instruments available," said Nicholas Tosca, a science-team

University of New York, Stony Brook. The first two newly accessed layers having sulfate salts and spherical concretions; both are signs of formatid conditions.

Squyres said, "I had thought we might see just basalt below the top saltyi as far as we've been able to see so far. Every time we see more sulfates it adds to the amount of water that was necessary to make this happen." JPL, a division of the California Institute of Technology in Pasadena, ma Rover project for NASA's Office of Space Science, Washington, D.C. Imageffi information about the project are available from JPL at http://marsroversj Cornell University, at http://athena.cornell.edu .

August 3, 2004

More Data From Mars Rover Spirit's First Month Now Online

Millions of people have viewed pictures from NASA's Spirit on the Mars rc Internet sites. Beginning today, a more complete set of science data from days is posted on a site primarily for scientists and technical research anyone who's interested.

The first installment of images, spectroscopic measurements, daily report from NASA's Mars Exploration Rover project has been posted on NASA's Plan available with a new "Analyst's Notebook" user interface at: http://pds-q; . Home page for the Planetary Data System is http://pds.:_pl.nasa.gov . In§ the system's Planetary Image Atlas, at http://pdsimg.lpl.nasa.gov/cqi-bin/MER/search?INSTRUMENT HOST NAME.MARS EXPLORATION ROVER . Data from Opi 30 martian days, or "sols," will be added Aug. 24, and data from later pcb missions will be added in October.

"All the raw images and selected processed images and other information Lei the public since the rovers first reached Mars in January. This release maps used for planning, all the non-image data from the spectrometers, dal and activity plans," said Dr. Ray Arvidson of Washington University, St. i investigator for the twin rovers' science payload.

"The 'Analyst's Notebook' is designed to help you navigate through the dab synergies," he said. "You can't deal with the Moessbauer spectrometer real without information about other observations that go with it."

"We are proud to be releasing such a comprehensive set of data from the s the twin rovers so quickly," said Dr. Jim Garvin, NASA's chief scientist the dedication and commitment of the science and engineering teams that tl collection of information is now available to the entire world for interpn help guide NASA's new exploration focus," added Garvin.

August 3, 2004

More Data From Mars Rover Spirit's First Month Now Online

Millions of people have viewed pictures from NASA's Spirit on the Mars rd, Internet sites. Beginning today, a more complete set of science data from days is posted on a site primarily for scientists and technical researchal anyone who's interested.

The first installment of images, spectroscopic measurements, daily report from NASA's Mars Exploration Rover project has been posted on NASA's Plan available with a new "Analyst's Notebook" user interface at: http://pds-qu . Home page for the Planetary Data System is http://pds.jpi.nasa.gov . In§ the system's Planetary Image Atlas, at http://pdsimq.jpl.nasa.gov/cgi-bin/MER/search?INSTRUMENT HOST NAME=MARS EXPLORATION ROVER . Data from Opi 30 martian days, or "sols7" will be added Aug. 24, and data from later pa's missions will be added in October.

"All the raw images and selected processed images and other information Lei the public since the rovers first reached Mars in January. This release as maps used for planning, all the non-image data from the spectrometers, dal

and activity plans," said Dr. Ray Arvidson of Washington University, St. i investigator for the twin rovers' science payload.

"The 'Analyst's Notebook' is designed to help you navigate through the dab synergies," he said. "You can't deal with the Moessbauer spectrometer real without information about other observations that go with it."

"We are proud to be releasing such a comprehensive set of data from the s the twin rovers so quickly," said Dr. Jim Garvin, NASA's chief scientist the dedication and commitment of the science and engineering teams that ti collection of information is now available to the entire world for interp help guide NASA's new exploration focUs," added Garvin.

Guy Webster (818) 354-6278

Jet Propulsion Laboratory, Pasadena, Calif.

Donald Savage (202) 358-1727

NASA Headquarters, Washington, D.0

News Release: 2004-194 August 5, 2004

Rocks Tell Stories in Reports of Spirit's First 90 Martian Days Scientific findings from the NASA rover Spirit's first three months on Ala marking the start of a flood of peer-reviewed discoveries in scientific jO two-rover adventure.

Researchers using Spirit's toolkit of geological instruments from early Ji record from rocks and soils in the rover's landing area and found a histo impact cratering, wind effects and possible past episodes of scant undergi Evidence for the water comes from mineral alteration in the veins, inclusj rocks. Eleven reports with 120 collaborating authors from around the work Aug. 6 issue of the journal Science.

"This is the first batch," said Dr. Steve Squyres of Cornell University, A for the science payload on both Mars Exploration Rovers. "You'll be seeilt months ahead and, no doubt, for many years to come based on information fm Opportunity. These machines just keep going and going, so the science jus coming." Dr. Jim Garvin, NASA's Chief Scientist for Mars added, "This is .8 remarkable scientific legacy of the rovers that will not only rewrite own also pave the way for human exploration."

The rovers completed three-month primary missions in April, then began ban extended science missions. "Spirit and Opportunity have really done yeom4 after more than twice as long as their original assignments. We don't knch keep working, but while they do we promise to keep them busy," said Jim g at NASA's Jet Propulsion Laboratory, Pasadena, Calif.

Both rovers were equipped and targeted to collect evidence about past env] especially any history of liquid water, since life as we know it depends inside Gusev Crater, an ancient Connecticut-sized impact basin that was an because it may have once held a giant lake fed by flows of water though into the crater.

The new reports state that, in its first three months, Spirit found no ev (lacustrine) deposits. "Any lacustrine sediments that may exist at this J apparently have been buried by lavas that have undergone subsequent impact leadoff paper by Squyres and 49 other rover science team members. Spiritk to a different location -- nearby hills over 3 kilometers (2 miles) awayg Dr. John Grant of the National Air and Space Museum, Washington, and co-ah rocks on the plain that Spirit explored during its primary mission incread maximum size as the rover got closer to an old 210-meter (690-foot-wide) that excavated the crater brought volcanic rocks to the surface from as d Several papers give evidence that rocks in the area are a volcanic type al mineral olivine. These include reports by Cornell's Dr. Jim Bell with col§ panoramic camera and by Dr. Dick Morris of NASA Johnson Space Center, HOUR using the Moessbauer spectrometer. Dr. Hap McSween of the University of T co-authors state, "These basalts extend the known range of rock compositij martian crust."

Dr. Ken Herkenhoff of Flagstaff, Ariz., offices of the U.S. Geological Sig Spirit's microscopic imager offer findings that rocks cut into by the rola

coatings and bright veins apparently from mineral alteration after the rol of Max-Planck-Insitut-fur-Chemie in Mainz, Germany, and other users of SS spectrometer report that bromine in the veins suggests the alteration res water. Dr. Phil Christensen of Arizona State University, Tempe, and collq miniature thermal emission spectrometer say the rock's coatings are consi moisture while buried. Dr. Ray Arvidson of Washington University, St. Loud cohesive texture in soils and rock coatings, which they suggest could re.± in the past.

Magnet experiments indicate almost all sampled dust particles in Mars' ath magnetic minerals, according to a paper by Dr. Preben Bertelsen of the Ni Copenhagen, Denmark, and others. Dr. Ron Greeley of Arizona State Universj that winds from the northwest grooved some rock surfaces and shaped sand They report that the way rock dust accumulates during grinding by Spirit'd that wind still comes from the same direction.

Guy Webster (818) 354-6278

Jet Propulsion Laboratory, Pasadena, Calif.

Donald Savage (202) 358-1727

NASA Headquarters, Washington, D.0

News Release: 2004-194 August 5, 2004

Rocks Tell Stories in Reports of Spirit's First 90 Martian Days Scientific findings from the NASA rover Spirit's first three months on Mad marking the start of a flood of peer-reviewed discoveries in scientific jj two-rover adventure.

Researchers using Spirit's toolkit of geological instruments from early record from rocks and soils in the rover's landing area and found a histai impact cratering, wind effects and possible past episodes of scant under Evidence for the water comes from mineral alteration in the veins, incluad rocks. Eleven reports with 120 collaborating authors from around the world Aug. 6 issue of the journal Science.

"This is the first batch," said Dr. Steve Squyres of Cornell University, A for the science payload on both Mars Exploration Rovers. "You'll be seeiryi months ahead and, no doubt, for many years to come based on information 40 Opportunity. These machines just keep going and going, so the science jus coming." Dr. Jim Garvin, NASA's Chief Scientist for Mars added, "This is .8 remarkable scientific legacy of the rovers that will not only rewrite our also pave the way for human exploration."

The rovers completed three-month primary missions in April, then began ban extended science missions. "Spirit and Opportunity have really done yeomaj after more than twice as long as their original assignments. We don't knob keep working, but while they do we promise to keep them busy," said Jim .g at NASA's Jet Propulsion Laboratory, Pasadena, Calif.

Both rovers were equipped and targeted to collect evidence about past envl especially any history of liquid water, since life as we know it depends inside Gusev Crater, an ancient Connecticut-sized impact basin that was because it may have once held a giant lake fed by flows of water though into the crater.

The new reports state that, in its first three months, Spirit found no ev (lacustrine) deposits. "Any lacustrine sediments that may exist at this lm apparently have been buried by lavas that have undergone subsequent impact leadoff paper by Squyres and 49 other rover science team members. Spiritk to a different location -- nearby hills over 3 kilometers (2 miles) awayg Dr. John Grant of the National Air and Space Museum, Washington, and co-ah rocks on the plain that Spirit explored during its primary mission incread maximum size as the rover got closer to an old 210-meter (690-foot-wide) that excavated the crater brought volcanic rocks to the surface from as d Several papers give evidence that rocks in the area are a volcanic type d mineral olivine. These include reports by Cornell's Dr. Jim Bell with col§

panoramic camera and by Dr. Dick Morris of NASA Johnson Space Center, Hot using the Moessbauer spectrometer. Dr. Hap McSween of the University of T co-authors state, "These basalts extend the known range of rock compositij martian crust."

Dr. Ken Herkenhoff of Flagstaff, Ariz., offices of the U.S.

Geological ag Spirit's microscopic imager offer findings that rocks cut into by the rola coatings and bright veins apparently from mineral alteration after the rol of Max-Planck-Insitut-fur-Chemie in Mainz, Germany, and other users of Sail spectrometer report that bromine in the veins suggests the alteration res water. Dr. Phil Christensen of Arizona State University, Tempe, and coil+ miniature thermal emission spectrometer say the rock's coatings are consi moisture while buried. Dr. Ray Arvidson of Washington University, St. Loud cohesive texture in soils and rock coatings, which they suggest could re± in the past.

Magnet experiments indicate almost all sampled dust particles in Mars' ath magnetic minerals, according to a paper by Dr. Preben Bertelsen of the Ni Copenhagen, Denmark, and others. Dr. Ron Greeley of Arizona State Universi that winds from the northwest grooved some rock surfaces and shaped sand i2 They report that the way rock dust accumulates during grinding by Spirit'd that wind still comes from the same direction.

JPL, a division of the California Institute of Technology in Pasadena, mag Rover project for NASA's Science Mission Directorate,

Washington. Images n about the project are available from JPL at http://marsrovers.lpl.nasa.go

Latest gNews From Spirit and Opportunity- September 22, 2004

http://marsrovers.ipl.masa.gov/mission/status spirit. html spirit on autopin place to continue daily science observations automatically throughout the conjuction period until sol 258.more.

http://marsrovers.jpl.nasa.gov/mission /status opportunity.html spectrometer and dirty targets opportunity took spectrometer readings on brushed areas sc called "Escher," including some areas cleaner than others. More>>

http://marsrovers.ipl.nasa.gov/mission/wir/ latest press image latest September 21

All Press Image <All Press Releases?

All raw images <Panorama images>[3-d image]

All videos

pis

The vision for space exploration

http://marsrovers.ipl.masa.gov/gallery/press/spirit/20040924a/mera-

[solo240A259R1](#) br. jpgsampling martian soil

Scientis were using the Mossbauer spectrometer on NASA'S mars explorating when something unexpected happen. The instruments' contact ring had beel onto the ground as a reference point of placement of another instruments particle X-rayspectrometer, for analyzing the soil. After spirit removed soil. The gap, about a centimeter wide (less than half an inch), is visib this mosaic of image more>>

Browse 227 MB large 1.7 mb

Opportunity's travels during its first 205 Martian days

[http://marsrovers.ipl.nasa.gov/gallery/press/opportunity/20040921a/enduran-b235r1br2.jpg](#) [browse] [medium](103 kb) large (386mg)

This map shows the traverse of NASA's mars Exploration Rover Opportunity rover's $205^{th}$ martian day, or sol (Aug. 21, 2004) The background image descent imaging camera Images inset along the route are from opportuniting camera. Opportunity began its exploration inside "Eagle" creater near the map. More>>

Look for museum events around the country. For an online experience visit marsquest online

http://marsrovers.ipl.nasa.gov/spotlight/index.html-missionfantasticmisnic

Bedrock in Mars' Gusev Crater Hints at Watery Past

Now that NASA's Mars Exploration Rover Spirit is finally examining bedroce finding evidence that water thoroughly altered some rocks in Mars' Gusev Spirit and its twin, Opportunity, completed successful three-month primal and are returning bonus results during extended missions. They remain in t beginning to show signs of wear.

On Opportunity, a tool for exposing the insides of rocks stopped workings optimistic that the most likely diagnosis is a problem that can be fixed pebble trapped between the cutting heads of the rock abrasion tool," said manager at NASA's Jet Propulsion Laboratory, Pasadena, Calif. "We think to the heads in reverse, but we are still evaluating the best approach to rar: are several options available to us."

Opportunity originally landed right beside exposed bedrock and promptly fb an ancient body of saltwater. On the other hand, it took Spirit half a yEi plain to reach bedrock in Gusev Crater. Now, Spirit's initial inspection a hill about 9 meters (30 feet) above the plain suggests that water may o Gusev.

"We have evidence that interaction with liquid water changed the compositib Steve Squyres of Cornell University, Ithaca, N.Y., principal investigator both rovers. "This is different from the rocks out on the plain, where we] apparently due to effects of a small amount of water. Here, we have a mom alteration, suggesting much more water."

Sc yres said, "To really understand the conditions that altered Clovis, wd like before the alteration. We have the 'after.' Now we want the 'before. be rocks nearby that will give us that."

Dr. Doug Ming, a rover science team member from NASA's Johnson Space Cent indications of water affecting Clovis come from analyzing the rock's surf alpha particle X-ray spectrometer and finding relatively high levels of Ldi inside the rock. He said, "This is also a very soft rock, not like the ba plains of Gusev Crater. It appears to be highly altered."

Rover team members described the golf-cart-sized robots' status and recei JPL today.

Opportunity has completed a transect through layers of rock exposed in thi stadium-sized "Endurance Crater." The rocks examined range from outcrops n through progressively older and older layers to the lowest accessible °Litt after a Canadian Arctic island. "We found different compositions in diffel Gellert, of Max-Planck-Institut fur Chemie, Mainz, Germany. Chlorine concn threefold in middle layers. Magnesium and sulfur declined nearly in paral8 layers, suggesting those two elements may have been dissolved and removed Small, gray stone spheres nicknamed "blueberries" are plentiful in Endura

Opportunity's smaller landing-site crater, "Eagle." Pictures from the rov a new variation on the blueberries throughout a reddish-tan slab called "i outcrop. "They're rougher textured, they vary more in size, and they're t instead of gray," said Zoe Learner, a science team collaborator from Corn" some cases where these are eroding, you can see a regular blueberry or ali One possibility is that a water-related process has added a coarser outer said, adding, "It's still really a mystery."

JPL, a division of the California Institute of Technology in Pasadena, me Rover project for NASA's Science Mission Directorate, Washington. Images zi about the project are available from JPL at http://marsrovers.lpl.nasa.gcth at http://athena.cornell.edu .

Guy Webster (818) 354-6278

Jet Propulsion Laboratory, Pasadena, Calif.

Donald Savage (202) 358-1547

NASA Headquarters, Washington, D.C.

NEWS RELEASE: 2004-234 September 21, 2004

Rover Missions Renewed as Mars Emerges From Behind Sun

As NASA's Spirit and Opportunity rovers resumed reliable contact with Earth, after a period when Mars passed nearly behind the Sun, the space agency extended funding for an additional six months of rover operations, as long as they keep working.

Both rovers successfully completed their primary three-month missions on the surface of Mars in April and have already added

about five months of bonus exploration during the first extension of their missions.

"Spirit and Opportunity appear ready to continue their remarkable adventures," said Andrew Dantzler, solar system division director at NASA Headquarters, Washington. "We're taking advantage of that good news by adding more support for the teamwork here on Earth that's necessary for operating the rovers."

Neither rover drove during a 12-day period this month, while radio transmissions were unreliable because of the Sun's position between the two planets. Daily planning and commanding of rover activities recommenced Monday for Opportunity and today for Spirit.

"It is a relief to get past this past couple of weeks," said Jim Erickson, project manager for both rovers at NASA's Jet Propulsion Laboratory, Pasadena, Calif. "Not only were communications disrupted, but the rovers were also going through the worst part of Mars southern-hemisphere winter from a solar-energy standpoint." "Although Spirit and Opportunity are well past warranty, they are showing few signs of wearing out," Erickson said. "We really don't know how long they will keep working, whether days or months. We will do our best to continue getting the maximum possible benefit from these great national resources."

Rover science team members will spend less time at JPL during the second mission extension. They are able to attend daily planning meetings by teleconferencing from their home institutions in several states and in Europe. "All 150 science team members and

collaborators have been provided the tools to be able to participate remotely," said JPL's Dr. John Callas, science manager for the rover project. Workstations researchers used at JPL are at their home institutions. Planning tools include video feeds, workstation display remote viewing, and audio conferencing.

Besides reducing costs, remote operations allow scientists to spend more time at home. "We get back to more normal lives, back to our families, and we still get to explore Mars every day," said Dr. Steve Squyres of Cornell University, Ithaca, N.Y., principal investigator.

Another change in operations is a shift from seven days per week to five days per week from October through December. This accommodates a temporary trim of about 20 percent in the project's engineering team to about 100 members. The rovers' reduced energy supply, during the rest of the martian winter, makes the inactive days valuable for recharging batteries. By January, the energy situation will have improved for the solar-powered rovers, provided they are still operating. The team size will rebound to support daily operations. As Mars emerges from behind the Sun, Spirit is partway up the west spur of highlands called the "Columbia Hills," a drive of more than 3 kilometers (2 miles) from its landing site. Opportunity is inside stadium-size "Endurance Crater," headed toward the base of a stack of exposed rock layers in "Burns Cliff," and a potential exit route on the crater's south side.

JPL, a division of the California Institute of Technology in Pasadena, manages the Mars Exploration Rover project for NASA's Science Mission Directorate, Washington. Images and additional information about the project are available on the Web at http://marsrovers.jpl.nasa.gov/ and http://athena.cornell.edu/ . For information about NASA programs on the Internet, visit

http://www.nasa.gov/.

NOW STORY ABOUT NASA ORBITER WHO IS MUCH OLDER THAN SPIRIT AND OPPURTUNIT CLOCKING MARTIAN SURFACE SOIL, AND WATER WHICH NOW IS CONFIRMED DOING THIS FOR ALMOST TWO YEARS OF EARTH AND ONE FULL MARTIAN YEAR.HAS CROSSED ITS ALLOTED TIME LIMIT, ITS GIVEN MORE AGE OVER MARS , BUT STILL EXPLORE A FUTURE AND SAFER WORLD FOR ALL OF US

RATNESH DWIVEDI SEPTEMBER 25 ,2004

Nasa jet propulsion laboratory manages the 2001 mars odyssey mission fort science,Washington d.c.the thermal imaging system was developed by Arizon university, tempe in collaboration with Raytheon santa, Barbara remote sens investigation is led by dr Philip cristenson at Arizona state university.1] aeronautics,Denver developed and built the orbiter.mission operations wash Lockheed martin and jet propulsion laboratory,Pasadena,calif.

http: //mail .yahoo. com/config/login?/http: //mail .yahoo. com/config/login?/.tt.htmlhttp://themis.la.asu.edu/mission.htmlhttp://themis.la.asu.edu/lates data.asu.edu/http://marsed.asu.edu/http://mars.lpl.nasa.gov/odyssey/

Ophir Chasma Etched Rock (Released 23 September 2004)

Note: this THEMIS visual image has not been radiometrically nor geometrici preliminary release. An empirical correction has been performed to remove linear shift has been applied in the cross-track and down-track directions and planetary motion. Fully calibrated and geometrically projected images, the Planetary Data System in accordance with Project policies at a later

| Lat: |
| --- |
| -15,165 |
| Lon: 315.002E |
| Click on link(s) below to view observation |
| details. |

20040923
A Pan
;Select ,Zoom
Where is Mars Odyssey Right Now?

The images below are all SIMULATED views. All of the images are computer automatically updated every 10 minutes.

http:/imars1.-
101.nasa.gov/mqs/realtime/odysse171.4pq

Mars As Seen From Earth

This image shows the position of the Mars Odyssey spacecraft relati http: /mars1.4pl.nasa.gov/mqs realtime odyssey.jpg

Looking Down On The Sun

This image shows the relative position of the planets in the inner

\Mars is relative to Earth.

;                                                                    ;
http://marsl.ipl.nasa.gov/mqs/realtime/odyssey3.jpq

Mars Odyssey's View of Earth

> Mars Odyssey was launched on April 7, 2001. The current distance indicated as "Range" in the box in the lower right hand corner_ of tye to M
>
> solar a
>
> the sppe grag:

Images above showes autogenerated images of earth,mars and odyssey orbiter. so far odysseyhas taken more than 1000 rounds and has comleted one full martian years orbiting red planet.it measured so far Where water and ice is buried on martian surface.Analysed what mars is made of by identifying minerals and chemicals and,Studied the martian radiation environment to help us understand potential health effects on future human explorersAs we celebrate Spirit's success, another of our robotic friends is celebrating an anniversary of sorts. Last week, NASA's Mars Odyssey orbiter reached an important milestone: a full Mars year (687 Earth days) of science mapping. During this martian year, it has: shown us where water ice lies buried beneath the surface;analyzed "what Mars is made of" by identifying minerals and chemical elements; and,studied the martian radiation environment to help us understand potential health effects on future human explorers."Before you send any landers to Mars, you want to look at the planet as a whole. We call that 'global reconnaissance,'" said Bob Mase, Odyssey Mission Manager. To help rovers like Spirit and Opportunity be successful, orbiters must do the crucial work of mapping the planet, identifying scientifically interesting locations and classifying potential hazards at the landing sites. Odyssey, like its predecessor, Mars Global Surveyor, is a valuable asset in the convoy of martian spacecraft NASA continues to send to the red planet.

"We are both limited in what we can do. Orbiters can't scrape rocks and look at them microscopically, and rovers cannot traverse and image the entire planet. So, the two types of missions really complement one another," Mase said, from the desk where he monitors the spacecraft. Beyond science studies of their own, orbiters have an important communications role to play. Not since Viking has NASA employed both orbiters and landed vehicles together

Today, the Odyssey and Mars Global Surveyor orbiters are helping the Spirit rover "talk" to ground controllers at JPL.
"It's difficult to communicate from the surface of Mars directly to Earth," Mase said. "You'd need a big antenna and a lot of power. It turns out that the rovers

can more efficiently send the information up to the orbiters, which are better equipped to relay the data back to Earth."

With this additional role, Odyssey team members may not have had much time to party on Odyssey's one-martian-year anniversary. They are receiving a pretty good gift though. During Spirit's time on Mars, 75% of the rover's pictures and data have come to Earth through Odyssey.

The Odyssey of Odyssey's 10,000 Orbits

steadfastly prepared for arrival at Mars and the grueling aerobraking process, which would drag the s; martian atmosphere to slow the spacecraft and bring Odyssey into, its desired mapping orbit.

"Every one of our 332 dives into the volatile martian atmosphere carried the potential to burn us up," back on it, the current success makes it easy to forget the pressure we were under. The eyes of the wt symbol of the American spirit, and the future of the Mars Program was on our shoulders as we were th to Mars following two failures. The notion that 'you cannot fail' still echoes in my mind," said Mase.Fail they did not, and are still making enormous contributions to current and future missions.

http://marsrovers.lpl.nasa.gov/newsroom/pressreleases/20040506a.htm1Panoramic Camera Mosaic tai

rover and transmitted to Earth via the Odyssey relay

Credit: NASA/JPL/Cornell

"The Mars Exploration Rover project wants to congratulate Odyssey on doing a tremendous job on bath own science investigations and at the same time enabling Spirit and Opportunity to make significant SC the surface of Mars," said Richard Cook, Mars Exploration Rover former project manager. "Almost all of coming down through the Odyssey relay, which is a testament to both the flexibility of the people on i capability of the design of the spacecraft."

"Odyssey's science return has been outstanding, and the entire team is eager to repeat this feat of 10,1 times in the years to come," said McSmith.

Learn more about Odyssey.

# ODYSSEY SCIENCE INSTRUMENT HIGHLIGHTS

The Thermal Emission Imaging System (THEM/St consists of two cameras: one that images visible wavele" infrared camera that is capable of detecting thermal and mineral variations. To date, THEM/S has mappe the infrared wavelengths and 15% in visible wavelengths. THEM/S images have been used to identify pc on martian hillsides, to find exposed water ice near the south pole, to provide the first complete high I south polar layered deposits, to map the remarkable region of Meridiani where the Opportunitiy rover ancient unweathered volcanic rocks and map unusual mineral deposits.

http://marsprogram.jpl.nasagov/spotlight/odyssey-10000-image07.htmlTHEMIS Visible Image of Spirit's

Crater

Credit: NASA/JPL/Arizona State University http://marsprogram]pl.nasa.gov/spotlight/odyssey-10000-image08.htmlGRS Thorium Elm Credit: NASA/JPL/University of Arizona

The Gamma Ray Spectrometer (CRS) is a suite of three instruments: a gamma subsystem located on the and two neutron subsystems, a neutron spectrometer and a high energy neutron detector. This collect collects gamma rays and neutrons emitted from the planet to determine the elemental composition of martian surface. The CRS has obtained full planet maps of the abundances of several elements including potassium, and thorium. CRS maps show very high contents of water ice buried just beneath the surfac polar regions. CRS has also measured the thickness of the annual carbon dioxide frost as the martian SE

http://marsprogram.ipl.nasa.gov/spotlight/oclyssey-10000-image09.htmlArtist's Concept of Future

Credit: NASA/JSC

The Martian Radiation Environment Experiment (MARIE) is a radiation monitor similar to those flown on the International Space Station. MARIE has measured the background radiation levels in orbit at Mars, as to 3 times that around the Earth. It has also served as an outpost for monitoring solar particle events.

Mission Success: The Magic of Mars Odyssey

Named after 2001: A Space Odyssey, the movie that inspired a generation to believe in a future where travelers on their way to Jupiter could call loved ones from space hotels via live television finks, NASA's 2001 Odyssey orbiter mission has actually brought that fantasy one step closer to reality -- via Mars.
With goals to detect health hazards for future human space explorers, to discover what our neighboring planet is made of, and to find buried water ice in the shallow subsurface of Mars, the 2001 Mars Odyssey orbiter has achieved mission success.

"As of August 24, 2004, the end date for its primary mission, Odyssey has officially fulfilled its science goals, and we look forward to refining our understanding of the red planet throughout an extended mission," said Dr. Jeff Plaut, Odyssey project scientist.
Artist's concept of future Humans on Mars. Image credit: NASA/JPL Link to Full

PAVING THE WAY FOR ASTRONAUTS TO MARS
One goal of the Odyssey mission was to analyze the radiation environment to determine its potential effects on human health. Future adventurers who rocket out

of the cradle of Earth towards Mars will leave Mother Earth's protective atmosphere and magnetic field.

"In order to build the best spacecraft to get humans to Mars and the safest habitats for humans to live in once they're there, we have to first know what exactly we're up against," explains Dr. Cary Zeitlin, principal investigator of the martian radiation environment experiment (MARIE).Protecting Humans from Health Hazards

Mars has less than one percent of the atmosphere of Earth and no magnetic shielding from solar flares and galactic cosmic radiation from outside the solar system. Since space radiation can cause cancer and damage to the central nervous systems of crew members on interplanetary missions, MARIE was created to calculate the radiation exposure humans would experience on the way to, and in orbit around, Mars. MARIE measured radiation levels and found that Mars radiation is about 2 to 3 times higher than that around Earth.

Mars radiantion levels are two to three times higher than around earth.
Image credit:
NASA/JPL/JSC/LAWRENCE BERKELEY NATIONAL LABORATORY
LINK TO FULL RES

Weather Satellites in Space 'For safe travel to Mars, we will need a coordinated system of satellites monitoring space weather at various locations throughout the solar system. The multiple satellites will help warn astronauts to go into 'storm shelters' that will protect them against intense, but relatively low-energy, solar flares."

The International Space Station (ISS) has well-shielded areas too, but in the lower Earth orbit of the /55, astronauts are still protected by the Earth's magnetic field. In contrast, during the deep space journey to Mars, astronauts could fly over 100 million miles outside of the 30,000-mile (50,000-kilometer) radius of the Earth's magnetic "shield." People living on Mars will likely need to limit the time they spend outside in their spacesuits as well as limit the distance they travel from their protective habitats. Astronauts on Mars will need to stay at a close enough distance to get to shelter when necessary. Dr. Cary Zeitlin, Principal Investigator for the Martian Radiation Environment Experiment Image credit : NASA/JPL Link to Full Res "Satellites with radiation detectors like MARIE will detect solar particles blasting out of the Sun early enough that astronauts could have about a half-hour's notice of an impending radiation storm," says Zeitlin.

"If the solar activity models get better, it might also be possible to predict particle events many hours or maybe days ahead of time. That would be a big help," Zeitlin adds.

If we send humans to Mars, Zeitlin believes that the potential benefit would be phenomenal. "There is so much people can do scientifically once they are on Mars, and the international cooperation that would flow out of putting that effort together would have big benefits on Earth," said Zeitlin.

When asked if he would want to go to Mars as an astronaut, Zeitlin responded, 7f I could come back to Earth."

## MARS STUDIES LEAD TO EARTH KNOWLEDGE

Home sweet home. "The more we learn about Mars, the more we learn about Earth," explained Dr. Bill Boynton, principal investigator for the gamma ray spectrometer (CRS) suite of instruments that is mapping the elemental makeup of Mars. "Whether it was taking apart a clock or a lawnmower engine in grade school, I've always had to understand how things work. For me, Mars is the biggest challenge because it seems to be the most complicated planet in our solar system, other than Earth itself," said Boynton.

What Earth and Mars are made of

Another goal of NASA's Mars Program and the Odyssey mission is to characterize the geology of Mars. CRS has allowed scientists to make maps of the elemental composition of the martian surface for the elements hydrogen, silicon, iron, potassium, thorium, and chlorine. These and other chemical elements are the building blocks of minerals, minerals are the building blocks of rocks, and all of these relate to the structure and landforms of the martian surface. An understanding of what Mars is made of in turn provides clues to the geological and climatic history of Mars and the potential for finding past or present fife.

"We see one element, potassium, which is about twice as abundant as it is on Earth," said Boynton. This fact was known in advance based on meteorite studies, but was confirmed by the CRS data. For the other elements, the team cannot yet tell if they are significantly different on Mars than on Earth.

Hydrogen, for example, is mainly tied up in water, but

scientists don't know the total amount on either Mars or Earth because they don't know how much is underground. "Even on Earth, there may be more water tied up in the Earth's mantle (between the surface and the core) than in all the water in the oceans," explained Boynton.

LIFE ON EARTH AND MARS

The elements Odyssey is finding on Mars are the same elements found on Earth. Some of these elements are essential to life as we know it. For example, iron is in our blood and helps carry oxygen to our lungs to breathe. Potassium is another trace element essential to fife. Finding these elements on Mars means that microbial life could have been (or still be!) present on Mars because they are chemical building blocks for life and its processes.

The chemical information gained from Odyssey provides a base of knowledge that scientists can build upon to determine the likelihood of fife on Mars.

This mosaic of day and night infrared THEWS images shows landslides flowed over 100 kilometers (62 miles) across the floor of Me/as Chasma.

Image credit: NASA/IPL/Arizona State University

Water is a key ingredient to life

On Earth, wherever there is liquid water, there is life. As far as biological research has shown, life seems to thrive in liquid, where molecules can run around and interact freely. Water, composed of hydrogen and oxygen, has a wide temperature range between freezing at 0 degrees Celsius (32 degrees Fahrenheit) and boiling at 100 degrees Celsius (212 degrees Fahrenheit), giving molecules a fair chance of combining and growing before locking up in frozen ice or boiling away as steam. What is unknown on Earth is how long it takes for life to form in liquid water. Odyssey and the Mars Program are attempting to find an answer.

How long does it take for life to develop in water?

"We don't know how long it takes for life to form in the presence of water," said Dr. Phil Christensen, principal investigator for the cameras on board

Mars Odyssey. Earth is a dynamic planet, dominated by water, with a lot of current volcanic activity, weathering, and plate-tectonic action. Both Earth and Mars are estimated at being about 4.5 billion years old. "The early history of Earth is so ground up due to current activity that you have to be very lucky to find rocks from the first half of Earth's history. But on Mars, which hasn't been weathered away, you may see rocks three or four billion years old right at the surface all over the planet," explained Christensen.

"So, ultimately, Mars may provide the best opportunity to figure out how an environment for supporting fife started on Earth."

> This mosaic of infrared mages hows the abundance and location of hematite at

WATER ON MARS...IN THE PAST

The thermal emission imaging system (THEMIS) on Odyssey is both an infrared camera and a visible camera It has captured telltale signs of past water on Mars. In four locations on Mars, THEMIS has detected high levels of hematite, a mineral that on Earth forms most often in the presence of liquid water.

Discoveries by THEMIS and its predecessor instrument(TES (thermal emission spectrometer) on the Mars Global Surveyor orbiter) led the 2003 Mars Exploration Rover mission team to choose Meridiani Planum as a landing site for its hematite content.

Since landing, data from the Opportunity rover's science instruments, including the miniature thermal emission spectrometer (mini-TES) has since confirmed THEMIS' results that the area was once covered in water.

The morphology and thermal properties of the Meridian! Planum region indicate that the hematite-bearing area was deposited in a standing body of water that extended over 100,000 square kilometers (300 miles by 100 miles or about the size of Oklahoma), with smaller bodies of water in nearby crater basins," said Christensen.

THEMIS, along with Christensen's two other instruments at Mars (TES on Mars Global Surveyor and mini-TES on the rovers), has shown that liquid water could have been in a few areas within the equatorial region of Mars for thousands or tens of thousands of years.

## ET ROVERS PHONE HOME THROUGH ODYSSEY

Beyond trying to solve the grand mysteries of life in the universe, Odyssey is chipping away at day-to-day tasks to help the rovers operate on Mars. Not only did the images from Odyssey and Mars Global Surveyor help plot the landing locations of the rovers, but Odyssey has also acted as a communications relay for the rovers, transmitting over 85 percent of the data from Spirit and Opportunity to Earth.

Pat Esposito, the lead navigator for the Mars Global Surveyor and Odyssey orbiters, explained that Odyssey's polar orbit takes the spacecraft over Spirit and Opportunity at least twice a day, once on the "day" side and once on the "night" side per rover. "Odyssey orbits Mars once every two hours, while Mars spins on its own axis. For every sequential orbit, Odyssey sees a different part of Mars underneath its cameras and instruments that is 28 degrees westward of its previous orbit ground-track," explained Esposito.

Odyssey has a much bigger antenna than the rovers have, so Odyssey can transmit more data faster to Earth, and the orbiter has Earth in view longer from its perch above Mars. Odyssey isn't quite sending streaming video from space like the movie 2001: A Space Odyssey predicted, but the orbiter is currently the exclusive data path for all the rover images. Almost nothing comes down to Earth that Odyssey doesn't deliver.

## SISTER SPACECRAFT IN A FAMILY OF MARS MISSIONS

Orbital images from Odyssey's THEMIS cameras have helped steer the twin rovers to their daily dream locations for geologic research. THEMIS can reveal what is underneath a uniformly dust-covered terrain, using its infrared camera that detects heat. Since rocks retain their heat through the cold nights on Mars, THEMIS can "see" where rocky terrain is versus a sandy area beneath an otherwise common looking surface. "Geologists want to study rocks more than dust because rocks tell you the environment they formed in unlike fine dust grains do," explained Christensen.

Odyssey is also already supporting the Phoenix Scout Mission with landing site selection images for Phoenix's scheduled landing on Mars in late May of 2008 in the north polar region. GRS has also detected significant quantities of water ice buried in the polar regions of Mars, which is a key motivation for sending Phoenix to this area.

hydrogen at the north and south poles. The areas shown in blue and violet are believed to consist of 50% water ice by volume.

Image credit: NASA/JPL/University of Arizona Link to Full Res

BURIED WATER ICE IN THE POLES

A final goal of the Odyssey mission was to search for subsurface water ice. "We have found that in the regions north and south of 60 degrees latitude, the surface is well over 50 percent water ice by volume.

If just the top meter of ice deposits around the martian north pole were melted, there would be enough liquid water to fill Lake Michigan," explained Boynton.

"We are now thinking that during a past climate there was enough moisture in the atmosphere that we actually had snow fall on the surface that was mixed with a small amount of wind-blown dust," said Boynton.

This discovery has significant implications for our understanding of the history of water and the possibilities for past and current fife on Mars as well as the potential for supporting human exploration of the planet, since humans will need water to survive on Mars.

Dr. Phil Christensen, Principal Investigator for THEM/S, works with students to enable them to take pictures of Mars with the cameras on Odyssey.

Image credit: NASA/JPL/Arizona State University

Link to Full Res

DEDICATED TEAM

The team behind the Odyssey mission is a dedicated group of individuals who have the passion to seek answers to difficult questions.

"I've just had a curiosity my whole life -- why certain mountains are in one place versus another, or why dinosaurs went extinct," said Christensen.

"As a scientist working on Mars, we get to ask similar questions about the formation of planets and the existence or non-existence of fife at a place where no humans have ever been - we get to see a world that has been untouched for 4.5 billion years."

Dr. Phil Varghese, Odyssey Project Manager.

Image credit: NASA/JPL Link to Full Res

END OF THE PRIMARY MISS/ON AND THE FUTURE OF ODYSSEY

August 24, 2004, marks the official end of Odyssey's picture-perfect primary science mission. Odyssey's new project manager, Phil Varghese, remembers watching the exciting evolution of the Odyssey mission as an "outsider" and admiring how launch, orbit insertion, and aerobraking all went perfectly. "Now, it's even more exciting to be directly involved in the mission as we reach the full mission-success criteria," said Varghese. "We have more than doubled the science return originally planned for

Odyssey, from 125 gigabits to over 250 gigabits of data."

The MARIE instrument on Odyssey was damaged last year, most likely by a solar particle entering the spacecraft and "frying" a computer chip. A similar incident happened during cruise, and after 6 months of remaining off, MARIE recovered itself. "We still have a glimmer of hope that it will turn back on 6 months from now since it recovered before by 'black magic,' which engineers officially call 'annealing," said Zeitlin.

Arthur C. Clarke, the author of 2001: A Space Odyssey, once said that, "Any sufficiently advanced technology is indistinguishable from magic." Built and operated by Lockheed Martin, the Odyssey spacecraft has definitely worked magic over the last 3 years.

"And Odyssey has enough flight system resources to carry on its science collection and relay capabilities through the next 10 years if things continue to go smoothly," said Varghese.

"My nieces and nephews ask for all kinds of information about Mars, and I have high hopes that we will continue our work and Mars student programs for many years to come."

More Instrument Information

For more detailed information about the instruments, please visit the instrument web sites.

The Gamma Ray Spectrometer, based at the University of Arizona, is a suite of instruments that includes a Neutron Spectrometer built by the Los Alamos National Laboratory and a High Energy Neutron Detector built by Russia's Space Research Institute.

The Martian Radiation Environment Experiment was built by Johnson Space Center and the Principal Investigator is based at Lawrence Berkeley National Laboratory at the University of California, Berkeley.

The Thermal Emission Imaging System is operated by the team at Arizona State University.

**My Mail To News Media after Mars Mission Bypassed alotted Time**

To: nik.gowing@bbc.co.uk, davidlessel@bbc.co.uk, lucy.hockingsObbc.co.uk   CC: ratneshdwivedi@indiainfacom

Subject: MARS ROVERS CROSSE ALLOTED TIME LIMIT----STILL PERFECT

Date: Thu, 26 Aug 2004 04:11:22 -1200

NASA ORBITER CREATES HISTORY, COMPLETES ONE MARTIAN YEAR ,GIVEN AGE, SILL PERFECT. Mars odyssey mission launched on april 7, 200lfinishes its given age by nasa scientists on mars.it has completed 687 dayes orbiting around mars.so far more than 10,000 orbital journey it has send themis images has explore water and ice and investigated radiation

health environment on mars to make it as second earth_an inhabbitent for future humns and explorers. This is yet a victorious step of nasa and sciences over time.

Ratnesh dwivedi   ratneshdwivedieindiainfo.com, ratniesh007@military.com

Ratnesh dwivedi is an internation journalist, a scientific researcher an has worked for bbc world wide, volunteered his services with white house on its policies.

----- Original Message

From: NASA Jet Propulsion Laboratory

Date: Wed, 25 Aug 2004 19:09:40 -0500

To: ratneshdwivedi@indiainfo.com

Subject: Mars Odyssey Begins Overtime After Successful Mission


Date: Wed, 25 Aug 2004 19:09:40 -0500

From: NASA Jet Propulsion Laboratory <info@jpl.nasa.gov>
Subject: Mars Odyssey Begins Overtime After Successful Mission
To:

ratneshdwivedieindiainfo.com

http://jpl.convio.net/site/R?iUtZMpaVIITuh0-3BCLCXxIg..Guy Webster (818) 354-6278 Jet Propulsion Laboratory, Pasadena, Calif.

Gretchen Cook-Anderson (202) 358-0836

NASA Headquarters, Washington, D.C.

RELEASE: 2004-209 August 25, 2004

Mars Odyssey Begins Overtime After Successful Mission

NASA's Mars Odyssey orbiter begins working overtime today after completing a prime mission that discovered vast supplies of frozen water, ran a safety check for future astronauts, and mapped surface textures and minerals all over Mars, among other feats. "Odyssey has accomplished all of its mission-success criteria," said Dr. Philip Varghese, project manager for Odyssey at NASA's Jet Propulsion Laboratory, Pasadena, Calif. The spacecraft has been examining Mars in detail since February 2002, more than a full Mars year of about 23 Earth months. NASA has approved an extended mission through September 2006.

"This extension gives us another martian year to build on what we have already learned," said JPL's Dr. Jeff Plaut, project scientist for Odyssey. "One goal is to look for climate change. During the prime mission we tracked dramatic seasonal changes, such as the comings and goings of polar ice, clouds and dust storms. Now, we have begun watching for year-to-year differences at the same time of year."

The extension will also continue Odyssey's support for other Mars missions. About 85 percent of images and other data from NASA's twin Mars rovers, Spirit and Opportunity, have reached Earth via communications relay by Odyssey, which receives transmissions from both rovers every day. The orbiter helped analyze potential landing sites for the rovers and is doing the same for NASA's Phoenix mission, scheduled to land on Mars in 2008. Plans call for Odyssey to aid NASA's Mars Reconnaissance Orbiter, due to reach Mars in March 2006, by monitoring atmospheric conditions during months when the newly arrived orbiter uses calculated dips into the atmosphere to alter its orbit into the desired shape.

Odyssey was launched April 7, 2001, and used the same dips into the atmosphere, known as aerobraking, to shape its orbit during the initial months after it reached Mars on Oct. 23, 2001. The spacecraft carries three research systems: a camera system made up of infrared and visible-light sensors; a spectrometer suite with a gamma ray spectrometer, a neutron spectrometer and a high-energy neutron detector; and a radiation environment detector. Less than a month after the science mapping campaign began, the team announced a major discovery. The gamma ray and neutron instruments detected copious hydrogen just under Mars' surface in the planet's south polar region. Researchers interpret the hydrogen as frozen water -- enough within about a meter (3 feet) of the surface, if the ice were melted,to fill Lake Michigan a couple times.

Here are a few of Odyssey's other important accomplishments so far:

-- As summer came to northern Mars and the north polar covering of frozen carbon dioxide shrank, Odyssey found abundant frozen water in the north, too.

-- Infrared mapping showed that a mineral called olivine is widespread. This indicated the environment has been quite dry, because water exposure alters olivine into other minerals. -- Findings indicated the amount of frozen water in some relatively warm regions on Mars is too great to be in equilibrium with the atmosphere, suggesting that Mars may be going through a period of climate change. Features visible near small, young gullies in some Odyssey images may be slowly melting snowpacks left over from a martian ice age.

-- The first experiment sent to Mars specifically in preparation for human missions found that radiation levels around Mars, from solar flares and cosmic rays, are two to three times higher than around Earth.

-- Odyssey's camera system obtained the most detailed complete global maps of Mars ever, with daytime and nighttime infrared images at a resolution of 100 meters (328 feet).

"We've accomplished everything we set out to do, and more," said JPL's Robert Mase, Odyssey mission manager. Although an unusually powerful solar flare in October 2003 knocked out the radiation environment instrument, Odyssey is otherwise in excellent health. The spacecraft has enough fuel onboard to keep operating through this decade and the next at current consumption rates. The mission extension, with a budget of $35 million, essentially doubles the science payoff from Odyssey for less than one-eighth of the mission's original $297 million cost.

JPL, a division of the California Institute of Technology, Pasadena, manages Mars Odyssey for NASA's Science Mission Directorate, Washington. Lockheed Martin Space Systems, Denver, built and operates the spacecraft. Investigators at Arizona State University, Tempe; University of Arizona, Tucson; NASA's Johnson Space Center, Houston; the Russian Aviation and Space Agency, Moscow; and Los Alamos National Laboratory, Los Alamos, N.M., built and operate Odyssey science instruments. For more information about Mars Odyssey on the Internet, visit: http://marsfpLnasagov/odyssey

http:7401.convio.net/site/R7i = ZsD4r2qY5e50-313CLCXxlg..Fronr NASA Jet Propulsion laboratory <infoejpLnasa.gov> (Save Addressl (Block Senderl (This Is Spam]

Show Full Headers

Previous / Next

http://jpl.convio.net/site/R?i- Obc5tfjtgFZ0-38CLCXxIg..Natalie Godwin (818) 354-0850

News Release: 2004-199 August 12, 2004

Mars Software Honored By NASA

NASA has selected a data visualization and simulation software package used by Mars rovers and landers, and a software package that can be used in aerospace and industrial flow fluid applications, as the "best of the best" software developed by the agency this year.

The "Science Activity Planner" developed by a team of experts at NASA's Jet Propulsion Laboratory, Pasadena, Calif, combines cutting-edge visualization with sophisticated planning and simulation capabilities to provide an intuitive interface to Mars rovers and landers. It is a multi-mission, multi-purpose tool that has achieved three simultaneous successes in mission operations, public outreach and technology development.

The software comes in two versions. The first is used in mission operations and contains the official mission activity dictionary. The second version was released under the name

"Maestro" to the public for education and outreach. Maestro includes additional training features that make it a more effective tool for public engagement.

The software is used heavily in the Mars Exploration Rover Mission on a daily basis. Scientists on the rover missions depend on the Science Activity Planner as their primary interface to the Spirit and Opportunity rovers. Every day, mission scientists and engineers use it to plan the next actions of the rovers and analyze the data arriving from Mars. The software has completed over 350 Mars days of successful mission operations of the Spirit and Opportunity rovers without a single critical failure and will continue to serve this role until the end of the mission.

"We are thrilled to receive this award and honored to have been a part of the amazing team behind the Mars Exploration Rover mission," said Jeff Norris, the software team leader at JPL. NASA also selected the TetrUSS 2004 software as an award winner. TetrUSS 2004 is a suite of computer programs used for fluid dynamics and aerodynamics analysis. Originally developed for NASA internal applications, TetrUSS 2004 has evolved into an efficient and versatile computer fluid dynamics tool used by engineers and scientists throughout the nation. The software is widely used in other government organizations, the aerospace industry, academia and non-aerospace industries such as automotive, biomedical and civil engineering.

Increased use of TetrUSS 2004 has occurred in critical NASA, government and industry programs. TetrUSS 2004 is now in use at over 500 sites for all classes of aerospace and industrial fluid flow applications, inside and outside of NASA, worth many billions of dollars. The TetrUSS 2004 team leader is Neal Frink of NASA's Langley Research Center, Hampton, Va. NASA began the competition in 1994, designed to reward outstanding software at the agency, as measured by the following criteria.

-- The science and technology significance of the software and its impact on NASA's mission -- The extent of current and potential use

-- The usability of the software

-- The quality factors considered in the software

-- Intellectual property factors such as patents and copyrights

-- Innovation of the software

Software eligible for this award must have NASA intellectual property interest, be of commercial grade, and be available to appropriate commercial users or dedicated to a NASA mission.

For more information about the Software of the Year award on the Internet, visit:

http://icb.nasa.gov/nasaswy.html

JPL, a division of the California Institute of Technology in Pasadena, manages the Mars Exploration Rover project for NASA's Science Mission Directorate, Washington. Images and additional information about the project are available from JPL at

http://marsrovers.jpl.nasa.gov and from Cornell University, Ithaca N.Y.

# CHAPTER -6: CASSINI HUYCENCE MISSIO N TO PROBE AND EXPLORE SATURNIAN SYSTEM

CASSINI HUYGENCE MISSION PROBE A JOINT VENTURE OF JET PROPULSION LABORATARY, EUROPEAN SPACE AGENCY AND ITALLIAN SPACE AGENCY, WAS LAUNCHED ON 15$^{TH}$ OCTOBER 1997 FROM NASA SPACE CENTRE IN FLORIDA.WITH MOTIVE TO EXPLORE SATURN,TS MYSTERIOUS RING S AND MANY MOONS . IT TRAVELLED BILIONS OF MILES OF JOURNEY THRU JUPITER'S ORBIT AND DEEP SPACE NETWORK,AND AFTER A SIX YEARS OF VOYAGE REACHED IN SATURNIAN SYSTEM IN THE MONTH OF JUNE/JULY 2004.

THE MISSION IS IMPORTANT TO ME ,BECAUSE I COOMUNICATED WITH JANE HUSTON JONES,SENIOR OUTREACH SPECIALIST WITH CASSINI MISSION,AND WITH WHOM I COMMUNICATED ABOUT MY DESIRE FOR A RESEARCH ON VARIOUS NASA MISSIONS AND SPACE SCIENC OPENED MY MIND ABOUT SPACE SCIENCE.MY THREE YEARS OF RESEARCH IS NOW SHAPED AS"THE COSMIC MASK".

ALSO SHE INTRODUCED ME WITH DR. LINDA SPIKER, WHO BEARS CREDIT TO BE ON FOUR PLANETARY MISSIONS .BOTH ARE MY ADVISERS. THE PROGRAMME WOULD BE CARRIED UNDER SUPERVISION OF WHITE HOUSE, NATIONAL RESEARCH COUNCIL/NATIONAL ACADEMICS WASHINGTON D.C. AND NASA CENTRES IN PASADENA AND FLORIDA.

RATNESH DWIVEDI SEPT 7 ,2004
BANGALORE. INDIA

First Things First: Initial Development
http://saturn.jp1.nasa.gov/cgi-bin/gs2.cgi?path =../multimedia/images/spacecraft/images/image2jpg&type = image
Technicians reposition and level the Cassini orbiter.

Before discussing the flight operations of the mission, an important element in the preparation of the spacecraft is the development phase. In this phase, development refers to the process of designing the spacecraft all the way from its conception through its final implementation.

After approval of the NASA proposal for Cassini by Congress several years ago, designers began the daunting task of devising plans and obtaining international cooperation for the Cassini orbiter and Huygens probe. Upon completion of the final design, engineers began to build the spacecraft. As fabrication progressed, it became necessary to test the newly completed assemblies.

The Cassini spacecraft's central computers, called the Command and Data Subsystem (CDS) and the Attitude and Articulation Control Subsystem (AACS) came to life" early on within their development. Their jobs were to replicate telemetry and command with the spacecraft through communication with the scientists, so that proper commands could be administered throughout Cassini's operation. This was accomplished in the form of the Test Telemetry and Command Subsystem (TTCS), a portable version of the ground system.

To accurately monitor and control the spacecraft while it remains in operation, scientists designed and assembled Cassini while simultaneously developing what is known as its ground system. The ground system, officially called Cassini Mission and Science Operations, is the collection of people, computers, software and procedures linking directly to the spacecraft.

Time lapse movie - building Cassini

QuickTime (2 MB)

QuickTime (4.4 MB)

MPEG MB)

MPEG MB)

This development phase will never really be complete, since new command sequences are continually being created throughout Cassini's operation. New flight software will also be developed and loaded aboard the spacecraft periodically, to run in the Command Data Subsystem,

Attitude and Articulation Control Subsystem, as well as in the physical science instruments onboard. The Spacecraft section contains detailed information about the subsystems and instruments. Flight software, as well as command sequences can be tested using duplicates of the Flight Command and Data Subsystem and Attitude and Articulation Control Subsystem within the Cassini Program's Integration and Test Laboratory. Rudimentary concepts outlined in Basics of Space Flight, an online reference, apply directly to the Cassini Huygens mission. Cassini Flight Operations ultimately means keeping track of the spacecraft as it seems to coast for years through the solar system. In reality, this means that engineers in Cassini Flight Operations are constantly busy-- regularly telling the spacecraft what to do, measuring its course and speed, making sure it's going where it's supposed to, as well as identifying and repairing any occasional glitches in the spacecraft or in the worldwide ground data system that keeps the spacecraft in touch with Earth

MISSION - Launch

http://saturn.jpl.nasa.gov/cgi-

Bin/gs2.cgi? path =./multimedia/images/launch/images/ image / jpg & type = imageCassini on the launch Standing majestically in Florida's early morning moonlit hours, the Titan IV-II/Centaur launch vehicle ca the Cassini-Huygens spacecraft was patiently awaiting the opening of the 140-minute launch window. days earlier, high winds prevented the launch. But on that morning, the 15th of October 1997, conditi were favorable.

Exactly 10 minutes before the opening of the launch window, the much-anticipated announcement ca through.

All systems are go!

Cheers followed the announcement. Hearts raced. This time, nature's elements were in sync with the l of the most sophisticated spacecraft ever built.

At the five-minute mark, and then again at the two-minute mark, an excited George Diller, Kennedy Sp Center's announcer, repeated the good news. "The status is 'go' across the board!"

At Launch Complex 40 on Cape Canaveral Air Station, the Mobile Service Tower retracted. The two upgi solid rocket motors mounted at the base prepared to blast the giant spacecraft out of Earth's atmospA Ten, nine, eight ...

http://saturn.jpl.nasa.gov/cgi-bin/gs2.cgi?path
=../multimedia/irnages/launch/images/imagel.jpg&type
imageCassini on the launch 1c Basking in the early morning fight, the imposing rocket stood proudly, as tall as a 22-story building. Ca wrapped inside the launch vehicle were the Cassini orbiter -- built by the Jet Propulsion Laboratory -- a Huygens probe, funded and built by the European Space Agency.

Six, five, four ...

After years of labor, tests, designs and redesigns, checks and double checks, more than 8,000 technich engineers in the United States -- and thousands more across 17 countries --

held their breaths, anxious the fruits of their labor go up in space.

Three, two, one ... Lift off!

Perfectly on schedule, at 4:43 a.m. EDT, the 5,650-kilogram (6-ton) spacecraft began its vertical ascent. by cheers from the crowd and the loud roars of the engines, furious flames propelled the spacecraft. matter of seconds, the full moon paled. Cassini-Huygens was indeed on its way toward Saturn. Scientis from every corner of the globe watched in glorious anticipation.

You can watch a video of the launch.

Two minutes and 23 seconds later, the announcer reported that the flawless launch sequence continue the separation from the Titan IV/B launch vehicle. By then, the spacecraft was already at an altitude of 91,440 meters (360,000 feet) and traveling at 7,046 kilometers (4,378 miles) per hour. http://saturn.jpl.nasa.gov/mission/images/cassini vk 2.jpgFamily and friends watch the launch.

That announcement received the loudest cheers and a few tears of joy in California at JPL's von Karma Auditorium, packed with family and friends gathered to witness the historical moment on the big scre even though it was nearly 2 A.M.

The Centaur upper stage separated successfully at 42 minutes and 40 seconds into the flight. Flying on own for the first time, 10 minutes later the Cassini-Huygens spacecraft successfully opened its communications fink with NASA's Deep Space Network tracking complex near Canberra, Australia

All systems on the spacecraft were operating normally, and data confirmed the precision of the launcl energy provided to the spacecraft by its launch vehicle was accurate to within one part in 5,000. At be than 0.04 degrees, any deviation in the trajectory (path) of the spacecraft was described as "insignific, http://saturn.jptnasa.gov/cgi-bin/gs2.cgi?path
=../multimedia/images/spacecraft/images/imagelajpg&type = imageDVD with over 6 signatures

Carrying a disk with signatures of more than 616,000 well-wishers from 81 nations, the Cassini-Huyge spacecraft was finally on its long journey to Saturn. Finally, the united efforts of three space agencie give humankind a sophisticated science laboratory in the orbit of the most fascinating planet in the s system. Once there, the spacecraft will collect data for four years.

With its majestic rings, dozens of frozen moons and a huge magnetosphere, Saturn has intrigued hum imagination for centuries. The planet's sheer distance from us created significant challenges, and whi spacecraft in the past two decades have succeeded in flying that far away from Earth, these voyages only appetizers in the voracious appetites of the science community.

http://saturn.jpLnasa.gov/cgi-

bin/gs2.cgi?path
=../multimediagmages/mission/images/image19.jpg&type= imageScientists' countrie: With Cassini-Huygens finally on its seven-year route toward the vast Saturnian system,

scientists could enjoy the main course. Starting in July 2004, one of the most sophisticated spacecraft ever sent into will bring them a nine-course meal: a feast of images and data that will answer many questions about mysterious corner of our solar system.

Truly an international enterprise, the mission enlisted help from 17 nations. Cassini was built and is in for NASA by the Jet Propulsion Laboratory. The European Space Agency contributed the Huygens Prob high-gain antenna and elements of several of Cassini's science instruments were provided by the !talk Space Agency. More than 260 scientists from 17 countries will examine the data.

Stay tuned. The best is yet to come.

CASSINI AT SATURN - Saturn Tour

assini's 6 Major Tour Segments

Saturn Orbit Insertion and Probe Release: Mite petal

Occultation Sequence:

grange Petal

Petal    Rotation & Magnetotail Petal:

1.dreen Petal :Titan 130 Transfer Jue Petal _Rotation/Icy
        Satellites:

, ellow Petal

High Inclination Sequence:

...ed Petal

rassini's landmark tour of Saturn begins July 1, 2004 UTC with the Saturn Orbit Insertion ISM engine burn. This ,turn will slow the spacecraft down, allowing it to be captured by Saturn's gravity. After the spacecraft is

^aptured, it will begin a 4-year tour of scientific exploration of the ringed planet, its moons, and magnetosphere.

The spacecraft will make 74 unique orbits around the planet, using 45 close flybys of Saturn's largest moon a itan for gravity assist and science data acquisition. Because of the sheer size of Titan, the flybys will allow '",r major changes in orbital paths, allowing engineers to minimize fuel use while maximizing science data ollection.

"assini's tour of the Saturn system is divided into 6 different segments. Each segment contains many Saturn and Titan flybys as well as opportunities to study the different smaller satellites.

˙ighlights of the Saturn Tour

/4 Orbits of Saturn

5 Close flybys of Titan

ta' close "targeted" flybys of other satellites:

- close flybys of Enceladus

.Phoebe

"yperion

Diane

'rea

Iapetus

bin/gs2.cgi?path =../multimedia/images/spacecraft/images/imagela jpg&type = imageDVD with over 6 signatures

Carrying a disk with signatures of more than 616,000 well- wishers from 81 nations, the Cassini-Huyge spacecraft was finally on its long journey to Saturn. Finally, the united efforts of three space agencie give humankind a sophisticated science laboratory in the orbit of the most fascinating planet in the s system. Once there, the spacecraft will collect data for four years.

With its majestic rings, dozens of frozen moons and a huge magnetosphere, Saturn has intrigued hum imagination for centuries. The planet's sheer distance from us created significant challenges, and whi spacecraft in the past two decades have succeeded in flying that far away from Earth, these voyages only appetizers in the voracious appetites of the science community.

http://saturn.jpl.nasa.gov/cgi-

bin/gs2.cgi?path=../multimediagmages/mission/images/image19.jpg&type= imageScientists' countrie: With Cassini-Huygens finally on its seven-year route toward the vast Saturnian system, scientists could enjoy the main course. Starting in July 2004, one of the most sophisticated spacecraft ever sent into will bring them a nine-course meal: a feast of images and data that will answer many questions about mysterious corner of our solar system.

Truly an international enterprise, the mission enlisted help from 17 nations. Cassini was built and is in for NASA by the Jet Propulsion Laboratory. The European Space Agency contributed the Huygens Prob high-gain antenna and elements of several of Cassini's science instruments were provided by the !talk Space Agency. More than 260 scientists from 17 countries will examine the data.

Stay tuned. The best is yet to come.


30 additional satellite flybys at distances less than 100,000 kilometers (about 62,100 miles)

Many Saturn and Ring occultation opportunities

One "Titan 180 degree" transfer

One high inclination sequence

For more information check out the Saturn tour schedule (PDF 187 KB) and details on the encounters with Saturn's moons (PDF 14 kE).

http://saturnjpl.nasa.gov/operations/images/cassini-tour Igipghttp://saturn.fpl.nasagov/operations/images/cassini-tour 2-Ig.jpgAbove is the Cassini-Huygens Mission Plorwith both aerial and edge-on views.

This plot outlines the tour of the spacecraft around Saturn, Titan and the icy satellites http://saturn.jpl.nasa.gov/operations/images/cassini-tour 2.1"Pg

Cassini Tour Subphase Boundaries

Segment
Start Date
End Date
Start Orbit
End Orbit
Subphase Title
1
-7/1/2004
2/15/2005
501
3
301 & Probe Release
2
2/15/2005
9/7/2005
3
14
3ccultation Sequence
3
:3/7/2005
7/22/2006
14
26

.petal Rotation / Magnetotail Petal
4
1/22/2006
6/30/2007
.16
47
180 Transfer
0/30/2007 V31/2007 47
.1g
flotation./ Icy Satellites
e/30/2007
"4/2008
49
-id of mission (741
Nigh Inclination Sequence

MISSION - Mission Control - Operations Overview
Mission Control is the heart of Flight Operations for unmanned, robotic spacecraft missions such as Galileo, Voyager, and Cassini-Huygens. All of the activities for tracking Cassini are controlled from within the Space Flight Operations Facility at JPL.

Cassini Ace Robert Springfield at Mission Control in JPL's Space Flight Operations Facility

One person is on site at all times when the Cassini spacecraft is being tracked in real time. This person is the mission controller, also called the "Ace." Depending on what activities are
scheduled for a particular day, there may be dozens of others involved with the Ace by
voice-net, telephone, or e-mail while they check system status, or send commands "up" to
the distant spacecraft. It is the Ace's job to ensure that all the spacecraft's data are
acquired, checked, stored, and distributed. That way, the engineers and scientists responsible for the spacecraft subsystems, the science experiments, and for navigation can
always access the data when they need to.

The computer workstations the Ace is watching in the above image show displays of the spacecraft's health and safety, as well as the real time status of the Deep Space Network WSN) antennas and systems that fink together to provide two-way communications with the spacecraft.

MISSION - Gravity Assists/Flybys

http://saturn.jpl.nasa.gov/mission/images/saturn-tilt.jpgSaturn

When the revolutionary Voyager-1 and Voyager-2 space missions launched in 1977 bound for Jupiter and 7aturn, their planned trajectories took them outward toward their target planets. Cassini was not able to follow this path.

:assini-Huygens is a massive spacecraft. It is carefully designed to brake into Saturn's orbit, as well as being loaded with an array of powerful instruments, cameras and sensors that will optimize the exploration of laturn's vast, distant system.

No existing launch vehicle could have sent the 6000-kg craft directly to Saturn.

the mission designers found that a technique called "gravity assist" was the answer. Gravity assist works because of the mutual gravitational pull between a moving planet and a spacecraft. The planet, of course, _mils on the spacecraft. Hut the spacecraft's own mass also pulls on the planet. This permits an exchange of _energy. For a detailed explanation of how the gravity assist technique works, you can read more about it in .:he Gravity Assist Primer.

http://saturn.jp1.nasa.gov/cgi-
Jin/gs2.cgi?path
../multimedia/videos/movies/trajechLmov&width     =
320&height 240&type = mCassini Trajectory Movie
juickTime WV

Cassini-Huygens has now looped around the Sun twice. On the first loop it flew dose behind Venus in its solar Jrbit, where it "stole" some of the planet's orbital momentum on April 26, 1998. The next loop provided a second flyby of Venus, on June 24, 1999, and one of Earth, on August 18, 1999. Given these three "gravity assist" boosts, Cassini-Huygens finally had enough orbital momentum to reach the outer solar system. One last - firavity assist from Jupiter on December 30, 2000 gave Cassini-Huygens the final thrust of energy it needed to ,project itself all the way to Saturn.

"rarity assist has often been called the "slingshot effect," but in reality, it is a different example altogether. Before letting go, a person wielding a slingshot whirls a projectile around and around, adding strength and,-efining aim each time, before letting go. The power comes from the thrower's muscles.

on the other hand, a spacecraft can obtain a gravity assist boost because the planet and the spacecraft tug 'Nn one another while orbiting the Sun. The spacecraft makes the planet lose some of its orbital momentum. rrom a planet's point of view, the spacecraft departs with no net energy gain, but from the Sun's point of ˉlew, the energy exchange makes a very small change in the planet's solar orbit. That amount of energy, though, benefits the tiny spacecraft substantially. The power comes from the planet's motion about the Sun.  ˉissuing Interplanetary Trajectory

once Cassini-Huygens reaches Saturn in July of 2004, the spacecraft will begin to fall toward the massive gas ˉfont. At just the right moment, Cassini will fire its main rocket engine for about 95 minutes to slow down. Instead of simply whizzing by, the spacecraft will become forever trapped in orbit like one of Saturn's moons.

The Huygens probe will separate from the Cassini orbiter and descend into Titan's murky atmosphere, while the orbiter examines Saturn's fascinating realm as it repeatedly loops around the planet.

MISSION - Navigation - Tracking Data
http://saturn.jpl.nasa.gov/mission/images/cassini-huygens.jpg
Cassini-Huygens Spacecraft

How do you keep track of an object that's been hurled away from Earth to travel for years throughout the solar system? How do you know where it is, and how fast it's traveling? The uplink and the downlink work together to solve these problems. Up/ink uses powerful radio transmitters, and downlink uses sensitive receivers, both within the Deep Space Network (DSN). Cassini-Huygens can only be tracked because it carries a radio transmitter that sends signals to Earth. (This is true with all other interplanetary spacecraft as well.) The transmitter aboard Cassini is linked with its own radio receiver, so that they can both work together when needed.

The two main types of tracking data that are used with Cassini are ranging and the Doppler effect. Using these two data types, the navigators can accurately track the Cassini-Huygens spacecraft.

Ranging

Ranging determines the distance (range) from Earth to the spacecraft and back, by placing specially coded signals (called ranging tones) on the radio up/ink, and recording the exact time as they go up. When the spacecraft receives them, it puts them on the downlink right away. When they come back to

Earth, the exact time is noted again. So basically, the ranging computer knows what time it sent the tones, and it knows what time they came back.

Since the speed of the radio signals is known (they travel at the speed of light), the round-trip distance can then be computed.

There are other factors to consider, too. How long did it take for the ranging tones to "turn around" inside the spacecraft's electronics? That miniscule delay is calculated from prelaunch testing. How long did it take the ranging tones to travel through the cable from the computer in the Deep Space Network (DSN) signal-processing center out to the radio telescope antenna before leaving Earth? The DSN finds that value while calibrating the system prior to each tracking period. And how far did the Earth move while the ranging pulses were traveling to the spacecraft? The navigators draw upon data gathered over years and years of observations by the astronomical community.

Highly evolved computer programs within the ranging system process these data to determine the distance between Earth and the Cassini-Huygens spacecraft.

Optical Navigation

Even though the most common means to track the spacecraft use ranging and the Doppler effect, a third data type can come into play once the spacecraft arrives near or is in orbit around Saturn. Optical navigation involves having the Cassini orbiter capture images of Saturn's satellites, with the background stars visible. These images come on the downlink as what is known as telemetry data, and once received, they are analyzed by the navigators for a more precise analysis of the spacecraft's trajectory than is available through ranging and the Doppler effect alone. Using this "opnav" data, instructions can then be up/inked to the spacecraft in the form of command data to Fine-tune the spacecraft's on-board schedule of science observations, or to fine-tune a direction to point its instruments.

All three of these types of data that Cassini-Huygens uses for navigation, are subject to the round-trip-fight time of around three hours across the distance between Earth and Saturn.

SCIENCE - Introduction Science Objectives QuickTime (1.6 MB) QuickTime (5.1 MB)

When considering such a complex, collaborative effort as the Cassini-Huygens mission, one might ask t

Study the interactions between the rings and Saturn's magnetosphere, ionosphere and atmosphere.

Study the interactions between the rings and Saturn's magnetosphere, ionosphere and atmosphere.

Titan Facts

Distance from Saturn:
1,221,850 kilometers = 759,220 miles
Orbital Period around Saturn:
15.945 days
Diameter:
5150 kilometers = 3200 miles (40% the diameter of Earth but still larger than Mercury and Pluto) Mass: $1.34 \times 10^{23}$ kilograms Average Density:
1.88 g/cm3
Temperature at surface: 95° Kelvin (178° C, -288° F)
SCIENCE - Why Study Titan?

http://saturnfpinasa.gov/cgi-bin/gsZcgi?path=../multimedia/images/titannmages/image18.jpg&type= image Titan is a time vault. It environment and thick atmosphere may resemble that of Earth some several billion years ago, before we know it began pumping oxygen into our atmosphere.

Titan is of great interest to scientists because it is the only moon known to have clouds and a mysteri planet-like atmosphere. Titan's atmosphere is made up primarily of nitrogen, which appears as an opad orange haze, obscuring its surface from view, and veiling its secrets. The Cassini-Huygens Mission seek that veil during its four-year tour of Saturn, its moons and magnetosphere. Scientists hope to gain a b understanding of Titan's surface, atmosphere and chemical composition, to perhaps shed some light a what primordial Earth might have been like billions of years ago. Titan is Saturn's largest moon, and was discovered by Christiaan Huygens in 1855_ Titan is the second la moon in the Solar System. Only Jupiter's moon Canymede is larger. At 5150 kilometers (3200 miles) in a Titan is larger than each of the planets Mercury and Pluto. Titan orbits Saturn at a distance of 9,222,0[ kilometers (759,478 miles), taking 15.9 days to complete one revolution.

SCIENCE - Icy Satellites

The science objectives of Saturn's icy satellites are as follows:

Saturn's Icy Satellites

Determine general characteristics and geological histories of Saturn's moons.

Define the different physical processes that have created the surfaces, crusts or sub-surfaces of the n Investigate compositions and distributions of surface materials, particularly dark, organic-rich materie condensed ices with low melting points.

Determine the bulk compositions and internal structures of the moons.

Investigate interactions of the moons with Saturn's magnetosphere and ring system.

The following table shows which instruments on the spacecraft support the corresponding science ob. above.

SCIENCE - The Formation of Saturn's Magnetosphere

Like Earth and Jupiter, Saturn's magnetic field is formed deep in the planet's interior. As the interior o cools, helium condenses in the liquid center of the planet. This condensation releases heat which pow convection in Saturn's interior. This convection powers the magnetic field.

Saturn's magnetic field is much weaker. than Jupiter's. Saturn does generate radio waves, but they are strong enough to be detected at Earth. Thus, until Pioneer 11 flew past Saturn in 1979, it was not knell whether Saturn even had a magnetic field. All we know about the magnetosphere surrounding Saturn

question; "Why go to Saturn in the first place?"

The answer comes even before the mission's conception. At the beginning, a basic set of science goals illustrated in a detailed, comprehensive mission plan that scientists hope to achieve and bring to the large.

Through the Cassini-Huygens Mission, we hope to gain a better understanding of the planet Saturn, its rings, its magnetosphere, its principal moon Titan and its other moons or "icy satellites."

Along with extensive preparation, planning and tracking throughout the mission, science objectives a divided into two parts: The goals that occur en route to the ringed planet, and then those that will oc the spacecraft arrives at the ringed planet.

En Route to Saturn:

http://saturn.jpl.nasa.gov/science/images/cassini-on-launch-padjpgCassini-Huygens on the launch pad To ensure all instruments onboard the spacecraft are working properly, checks were performed via re control at launch time as well as after 14 months in space. This includes routine instrument maintenar, Gravitational Wave Experiments, instrument calibrations, and conjunction experiments for a time perm least 30 days. These checks are performed to prepare Cassini for its upcoming tour of Saturn.

There are 12 instruments onboard the Cassini Spacecraft orbiter, and 6 instruments onboard the Huyg Probe. These instruments are all designed to perform in-situ Ion-sitel studies of elements of Saturn, it atmosphere, moons, rings and magnetosphere. The instruments will study various temperatures in va, locations, plasma levels, neutral and charged particles, compositions of surfaces, atmospheres and rim wind, and even dust grains in the Saturn system. Other instruments will perform spectral mapping for quality images of the ringed planet, its moons and rings.

The sophistication of the design and implementation of the instruments onboard the spacecraft creat other benefits, including the potential for technology spin-offs into the commercial world, internatioi cooperation and education and discovery for people of all ages.

This section illustrates the important science objectives for the Cassini-Huygens mission, which have 51 the crucial impetus for going to Saturn in the first place.

SCIENCE - Saturn

Saturn's science objectives are as follows:

Saturn

Determine the temperature field, cloud properties and composition of Saturn's atmosphere.

Measure the planet's global wind field, including its waves; make long-term observations of cloud feat see how they grow, evolve and dissipate.

Determine the internal structure and rotation of the deep atmosphere.

Study daily variations and relationship between the ionosphere and the planet's magnetic field. Determine the composition, heat flux and radiation environment present during Saturn's formation al evolution.

Investigate sources and nature of Saturn's lightning.

SCIENCE - Rings

The science objectives of Saturn's mysterious rings are as follows:

http://saturnfpl.nasa.gov/science/images/rings1208browsejpg

Saturn's Rings

Study configuration of the rings and dynamic processes responsible for ring structure. Map the composition and size distribution of ring material.

Investigate the interrelation of Saturn's rings and moons, including imbedded moons. Determine the distribution of dust and meteoroid distribution in the vicinity of the rings.

been learned through the combined encounters by Pioneer, Voyager 1, and Voyager 2, with some adc contributions by Hubble.

So where does Saturn's magnetic field originate? Scientists are not sure. It could have originated dee, core of the planet. Or it could have been generated further out in the planet. The only thing scientist that Saturn's extensive atmosphere has some effect in its formation.

Lining It All Up

Planetary magnetic fields like Earth's and Saturn's can be approximated by a dipole, a simple magneti structure with a north and sole pole. A bar magnet is one example of a dipole. Magnetic field measure from Pioneer and Voyager revealed a dipole-like field at Saturn with no measurable tilt between Satu rotation and magnetic axes. This near-perfect alignment of the two axes is unique among the planets comparison, Earth's dipole tilt is 11.4 degrees and Jupiter's is 9.6 degrees while Saturn's is less than oe degree.

Although Saturn's magnetic field is symmetrical about the planet's rotation axis, a number of the On processes observed by Pioneer and Voyager can only be explained by an asymmetric magnetic field reminiscent of Earth's. These observed phenomena include major radio emissions from Saturn (called , kilometric radiation) and the formation of ring spokes in the B-Ring.

A planetary magnetosphere forms when the solar wind (the supersonic, ionized gas that flows from t encounters a planet possessing a large magnetic field. The magnetic field forms a shield around the p forcing the solar wind to flow around

the magnetosphere. The region inside the magnetosphere is pc by the plasma generated from the magnetic field while outside the magnetosphere, the solar wind is dominant plasma
NOW ,1TS HAWBEEN RECEIVED WITH JPL ABOUT _CASSIN'S ARRIVAL 0111
SATURN,COMMUNICATIONS WITH JANE HUSTON JONES ON THE MISSION.-
kello ratnesh dwivedi,
Wednesday June 23, at 10 am. PDT, NASA TV will broadcast a press conference on preliminary results from the Phoebe flyby. Check the Cassini website for a webcast or press release if you don't get NASA TV.
Dr Linda Spilker will be the next Cassini speaker Tuesday June 29. 12:30 p.m. Pacific time, Toll-free in USA, 888-323-4924 PASSCODE: MUSEUM The telecon transcript will be available a week or so after the talk.
Dr. Spilker has held key roles on major space science missions that have explored four of the nine planets and their many moons. She is currently the deputy project scientist on the Cassini Mission to Saturn. She'll set the scene for Saturn Orbit Insertion and much more! Read more about Dr. Linda Spilker here:
http://ssajOnasa.gov/people/profile.cfm?Code= SpilkerL
On June 30, the live commentary from JPL for Saturn Orbit Insertion will begin at 6:30 p.m. PDT and continue through 9.˙45 p.m. The confirmation that Cassini has achieved Saturn orbit will be about 9:40 p.m. After a short break, there will be a press briefing about 10 p.m. First pictures the next

morning about 10 a.m. PDT. Keep an eye on http://www.nasa.gov/multimedia/nasatv/MM NTV Breaking.html for schedule changes.
The Press Kit for Saturn Arrival is now available at http://saturn/index.cfm Jane
Jane Houston Jones
Senior Outreach Specialist, Cassini Program
JPL - 4800 Oak Grove Drive, MS 230-205
Pasadena, CA 91109

Top of Form
Bottom of Form Printable Version Flag this message.
From: "ratnesh dwivedi" <ratneshdwivedieindiainfo.com> (Save Address) (Block Sender]
(This Is Spam]
Top of Form
To: jane.h.jonesPjpl.nasa.gov
CC:
Subject: LETTER TO A SPACE SCIENTIST
Date: Fri 25 Jun 2004 22:21:59 # 0530
Bottom of Form Show Full Headers Previous / Next Bottom of Form dear Jane,

i do not have words to express my glad ness to recieve info from your side and jpLactually i
have a depth interest in space science.I have written a 70 pages on universe our planets
thier origin and life possibilities on them.i wish to send it across u. will u give a littletime over
that.
may be my dream to work with jpl-nasa as aresearcher come true.' hold a biology science
degree from india and masterd in english journalism.aS AJOURNALIST I HAVE WORKED WITH MC WORLD.HENCE FORTH I WISH TO ASK U IF I SEND THE INFO PROVIDED BY U TO BBC WORLD AND OTHER MEDIA ORGANISATION.
ABOVE ALL ON PRESIIDENT'S OFFICIAL WEBADDRESS.ALSO I WOULD BE GLAD TO GET A JOB CHANCE
WITH CASSIN -HUGENS MISSION.ALREADY PUTTING HOURS WITH ONLINE ADDRESSES OF JPL-NASA.
DO REC/EVE MY DETAILED RESUME.
WARM REGARDS RATNESHDW/VEDI
   ---riginal Message
From: jane.h.jones@jpl.nasa.gov
Date: Tue, 22 Jun 2004 17:39:46 -0700
To: ratneshdwivedi@indiainfo.com
Subject: Press briefing June 23 and 30, Telecon June 29
Hello ratnesh dwivedi,

Wednesday June 23, at 10 a.m. PDT, NASA TV will broadcast a press conference on preliminary results from the Phoebe flyby. Check the Cassini website for a webcast or press release if you don't get NASA TV.

Dr Linda Spilker will be the next Cassini speaker Tuesday June 29. 12:30 p.m. Pacific time, Toll-free in USA, 888-323-4924 PASSCODE: MUSEUM The telecon transcript will be available a week or so after the talk.

Dr. Spilker has held key roles on major space science missions that have explored four of the nine planets and their many moons. She is currently the deputy project scientist on the Cassini Mission to Saturn. She'll set the scene for Saturn Orbit Insertion and much more!

Read more about Dr. Linda Spilker here: http://sse.jpl.nasa.gov/people/profile.cfm?Codez-SpilkerL

On June 30, the live commentary from JPL for Saturn Orbit Insertion will begin at 6:30 p.m. PDT and continue through 9.45 p.m. The confirmation that Cassini has achieved Saturn orbit will be about 9.40 p.m. After a short break, there will be a press briefing about 10 p.m. First pictures the next morning about 10 a.m. PDT. Keep an eye on http://www.nasa.gov/multimedia/nasatv/MM NTV Breaking.html for schedule changes.

The Press Kit for Saturn Arrival is now available at
http://saturn/index.cfm
Jane
Jane Houston Jones
Senior Outreach Specialist, Cassini Program
JPL - 4800 Oak Grove Drive, MS 230-205
Pasadena, CA 91109 Phone - 818-393-6435 Fax - 818-393-4495
jane.h.jonesPjpinasa.gov
http://saturnjpl.nasa.gov

Thank you for your resume. I am in the Cassini Outreach department, not the personnel department. To apply for a job at NASA JPL, you must use the Employment website http://careerlaunch.jpl.nasa.gov/

dear Jane,

this is one old and detailed resume of me,i am sending to u for a job of researcher/or else.keeping the current and future prospects.my address is changed and it is in my earlier mail to u .

i have alredy started enjoying images from cassini.the images of phobeas which acciorsing to cassini's info is made up of ice of distant past.

cassini is still 4dayes/ 4hours/ afar.

with in next few dayes u would recieve my article so u wold be able to understand

the way i think.

warm regards

ratneshdwivedi

bang/ore

# 9108025287140

Jane Houston Jones
Senior Outreach Specialist, Cassini Program JPL - 4800 Oak Grove Drive, MS 230-205 Pasadena, CA 91109
Phone - 818-393-6435
Fax - 818-393-4495
janah.jonesejpLnasa.gov
http://saturn.jpLnasa.gov
Carolina Martinez (8181354-9382
Jet Propulsion Laboratory, Pasadena, Calif. Donald Savage (202) 358-1727
NASA Headquarters, Washington NEWS RELEASE: 2004-158 June 23, 2004 Cassini Opens A Cosmic Time Capsule

Like a woolly mammoth trapped in Arctic ice, Saturn's small moon Phoebe may be a frozen artifact of a bygone era, some four billion years ago. The finding is suggested by new data from the Cassini spacecraft.

Cassini scientists reviewed data from the spacecraft's June 11, 2004, flyby of the

diminutive moon. They concluded Phoebe is likely a primordial mixture of ice, rock and carbon-containing compounds similar in many ways to material seen in Pluto and Neptune's moon Triton. Scientists believe bodies like Phoebe were plentiful in the outer reaches of the solar system about four and a half billion years ago.

These icy planetesimals (small bodies) formed the building blocks of the outer solar system and some were incorporated into the giant planets Jupiter, Saturn, Uranus and Neptune. During this process, gravitational interactions ejected much of this material to distant orbits, joining a native population of similar bodies to form the Kuiper Belt.

"Phoebe apparently stayed behind, trapped in orbit about the young Saturn, waiting eons for its secrets to be revealed during its rendezvous with the Cassini spacecraft," said Dr. Torrence Johnson, Cassini imaging team member at NASA's Jet Propulsion Laboratory, Pasadena, Calif.

"All our evidence leads us to conclude, Phoebe's surface is made of water ice, water-bearing minerals, carbon dioxide, possible clays and primitive organic chemicals in patches at different locations on the surface," said Dr. Roger N. Clark, team member for the visual and infrared mapping spectrometer, U.S. Geological Survey in Denver. "We also see spectral signatures of materials we have not yet identified." Cassini's observations gave scientists the first detailed look at one of these primitive icy planetesimals.

"One intriguing result is the discovery of possible chemical similarities between the materials on Phoebe and those seen on comets," said Dr. Robert H. Brown, team leader ▸ or the visible and infrared mapping spectrometer, University of Arizona, Tucson. Evidence that Phoebe might be chemically kin to comets strengthens the case that it is similar to Kuiper Belt Objects.

Measurements taken by the composite infrared spectrometer were used to generate temperature maps. The maps show the surface of Phoebe is very cold, only about 110 degrees above absolute zero (minus 163 degrees Celsius, or minus 261 degrees Fahrenheit). Even colder nighttime temperatures suggest a fluffy, porous surface layer.

"One of the first results from this map is the surface of Phoebe has been badly chewed up, probably by meteorite impacts," said Dr. John Pearl, a Cassini co-investigator for the composite infrared spectrometer, at NASA's Goddard Space Flight Center, Greenbelt, Md. "We are discovering Phoebe is a very complex object, with large variations in topography." Cassini also made radar observations of Phoebe's enigmatic surface, making it the first spacecraft radar observations of an outer-planet moon. The results are consistent with the dirty, rocky, icy surface suggested by other observations.

"We have conducted our first analysis of an outer solar system resident akin to Kuiper Belt Objects," said Dr. Dennis Matson, project scientist of the Cassini-Huygens mission at JPL. "In two short weeks, we have added more to what we know about Phoebe than we had learned about it since it was discovered 100 years ago. We did this by having multiple instruments conducting investigations all at one time during our flyby."

The Cassini-Huygens mission is a cooperative project of NASA, the European Space Agency and the Italian Space Agency. JPL manages the mission for NASA's Office of Space Science, Washington. For the latest images and more information about the mission on the Internet, visit http://www.nasa.gov and http://saturn.jpl.nasa.gov

Image Advisory: 2004-162 June 25, 2004

Getting Closer To Titan

Irregular bright and dark regions of yet unidentified composition and character are becoming increasingly visible on Titan's surface as Cassini approaches its scheduled first flyby of Saturn's largest moon on July 2, 2004.

This view represents an improvement in resolution of nearly three times over the previous Cassini images of Titan. Titan's surface is difficult to study, veiled by a dense hydrocarbon haze that forms in the high stratosphere as methane is destroyed by sunlight. This image is different from previous Titan images by Cassini because it was taken through a special filter, called a polarizer, which is designed to see through the atmosphere to the surface. Cassini will conduct a critical 96-minute burn before going into orbit around Saturn on June 30 (July 1 Universal Time), with its first

scheduled flyby of Titan on July 2.

Cassini Spacecraft Arrives At Saturn

The international Cassini-Huygens mission has successfully entered orbit around Saturn. At 9:12 p.m. PDT on Wednesday, flight controllers received confirmation that Cassini had completed the engine burn needed to place the spacecraft into the correct orbit. This begins a four-year study of the giant planet, its majestic rings and 31 known moons.

"This is a tribute to the team at NASA and our partners at the European Space Agency and the Italian Space Agency, to accomplish this feat taking place 934 million miles 11.5 billion kilometersl away from Earth," said Dr. Ed Weiler, associate administrator for space science at NASA Headquarters, Washington, D.C. "What Cassini-Huygens will reveal during its tour of Saturn and its many moons, including Titan, will astonish scientists and the public. Everyone is invited to come along for the ride and see all this as it is happening. It truly is a voyage of discovery."

Members of the Cassini-Huygens mission at NASA's Jet Propulsion Laboratory, Pasadena, Calif., broke into cheers and high-fives as NASA's Deep Space Network confirmed receipt of the signal indicating successful entry into orbit.

"We didn't expect anything less and couldn't have asked for anything more from the spacecraft and the team," said Robert T. Mitchell, program manager for the Cassini-Huygens mission at JPL. "This speaks volumes to the tremendous team that made it all happen."

Dr. Charles Elachi, JPL director and team leader on the radar instrument onboard Cassini, said, "It feels awfully good to be in orbit around the lord of the rings. This is the result of 22 years of effort, of commitment, of ingenuity, and that's what exploration is all about."

The mission will face another dramatic challenge in December, when the spacecraft will release the piggybacked Huygens probe - provided by the European

Space Agency - which will plunge through the hazy atmosphere of Saturn's largest moon, Titan.

"This was America's night. This was NASA doing it right," said Dr. David South wood, director of scientific programs for the European Space Agency. "They really gave those of us in Europe a challenge. We've got six months to go until we land on Titan. We're just praying that everything will go as well."

Saturn is the sixth planet from the Sun. It is the second largest planet in our solar system, after Jupiter. The planet and ring system serve as a miniature model of the disc of gas and dust surrounding our early Sun that eventually formed the planets. Detailed knowledge of the dynamics of interactions among Saturn's elaborate rings and numerous moons will provide valuable data for understanding how each of the solar system's planets evolved.

Cassini traveled nearly 3.5 billion kilometers (2.2 billion miles) to reach Saturn after its launch from Cape Canaveral Air Force Station, Fla., on Oct. 15, 1997. During Cassini's four-year mission, it will execute 52 dose encounters with seven of Saturn's 31 known moons. The first images are expected to return Thursday morning. Science measurements gathered Wednesday are the closest ever obtained of Saturn. Those measurements may reveal

details of the gravitational and magnetic fields that tell scientists about Saturn's interior. The Cassini-Huygens mission is a cooperative project of NASA, the

European Space Agency and the Italian Space Agency. The Jet Propulsion Laboratory, a division of the California Institute of Technology in Pasadena, manages the Cassini-Huygens mission for NASA's Office of Space Science, Washington, D.C. JPL designed, developed and assembled the Cassini orbiter.

For the latest images and more information about the Cassini-Huygens mission, visit http://saturn.jpl.nasa.gov and http://www.nasa.gov/cassini

July 1, 2004

Fresh Cassini Pictures Show Majesty of Saturn's Rings

The first pictures taken by the Cassini spacecraft after it began
orbiting Saturn show breathtaking detail of Saturn's rings, and

other science measurements reveal that Saturn's magnetic field pulsed in size as Cassini approached the planet.

"For years, we've dreamed about getting pictures like this After all the planning, waiting and worrying, just seeing these first images makes it all worthwhile," said Dr. Charles Elachi, Cassini radar team leader and director of NASA's Jet Propulsion Laboratory, Pasadena, Calif. "We're eager to share these new views and the exciting discoveries ahead with people around the world."

The narrow angle camera on Cassini took 61 images soon after the main engine burn that put Cassini into orbit on Wednesday night. The spacecraft was hurtling at 15 kilometers per second (about 34,000 miles per hour), so only pieces of the rings were targeted.

"We won't see the whole puzzle, only pieces, but what we are seeing is dramatic," said Dr. Carolyn Porco, Cassini imaging team leader, Space Science Institute, Boulder, Colo. "The images are mind-boggling, just mind-boggling. I've been working on this mission for 14 years and I shouldn't be surprised, but it is remarkable how startling it is to see these images for the first time."

Some images show patterned density waves in the rings, resembling stripes of varying width. Another shows a ring's scalloped edge.

"We do not see individual particles but a collection of particles, like a traffic jam on a highway," Porco said. "We see a bunch of particles together, then it clears up, then there's traffic again." Other instruments on Cassini besides the camera have also been busy collecting data. The magnetospheric imaging instrument took the first image of Saturn's magnetosphere. "With Voyager we nferred what it looked like, in the same way that a blind man feels an elephant. Now we can see the elephant," said Dr. Tom Krimigis of Johns Hopkins Applied Physics Laboratory, Laurel, Md., principal investigator for the magnetospheric imaging instrument. The magnetosphere is a bubble of energetic particles around the planet shaped by Saturn's magnetic field and surrounded by the solar wind of particles speeding outward from the Sun.

"During approach to Saturn, Cassini was greeted at the gate," said Dr. Bill Kurth, deputy principal investigator for the radio and plasma wave science instrument onboard Cassini. "The bow shock where the solar wind piles into the planet's magnetosphere was encountered earlier than expected. It was as if Saturn's county line had been redrawn, and that was a surprise." Cassini first crossed the bow shock about 3 million kilometers (1.9million miles) from Saturn, which is about 50 percent farther from the planet than had been detected by the Pioneer, Voyager 1 and Voyager 2 spacecraft that flew past Saturn in 1979, 1980 and 1981.

The location of the bow shock varies with how hard the solar wind is blowing, Kurth said. As the magnetosphere repeatedly expanded and contracted while Cassini was approaching Saturn, the spacecraft crossed the bow shock seven times.

Cassini Provides New Views of Titan, Saturn's Largest Moon The Cassini spacecraft has revealed surface details of Saturn's moon Titan and imaged a huge cloud of gas surrounding the planet-sized moon.

Cassini gathered data before and during a distant flyby of the orange moon yesterday. Titan's dense atmosphere is opaque at most wavelengths, but the spacecraft captured some surface details, including a possible crater, through wavelengths in which the atmosphere is clear.

Although the initial images appear bland and hard to interpret, we're happy to report that, with a combination of instruments, we have indeed seen Titan's surface with unprecedented clarity. We also look forward to future, much closer flybys and use of radar for much greater levels of surface detail," said Dr. Dennis Matson of NASA's Jet Propulsion Laboratory, Pasadena, Calif., project scientist for the international Cassini-Huygens mission. Cassini's visible and infrared mapping spectrometer pierced the smog that enshrouds Titan. This instrument, capable of mapping mineral and chemical features of the moon, reveals an
exotic surface bearing a variety of materials in the south and a circular feature that may be a crater in the north. Near-infrared colors, some three times redder than the human eye can see, reveal the surface with unusual clarity.
"At some wavelengths, we see dark regions of relatively pure water ice and brighter regions with a much higher amount of non-ice materials, such as simple hydrocarbons. This is different from what we expected. It's preliminary, but it may change the way we interpret fight and dark areas on Titan," said JPL's Dr. Kevin Baines, Cassini science-team member. A methane cloud is visible near the south pole. It's made of unusually large particles compared to the typical haze particles surrounding the moon, suggesting a dynamically active atmosphere there."
This is the first time scientists are able to map the mineralogy of Titan. Using hundreds of wavelengths, many of which have never been used in Titan imaging before, they are

creating a global map showing distributions of hydrocarbon-rich regions and areas of icy material. Cassini's camera also sees through the haze in some wavelengths. "We're seeing a totally alien surface," said Dr. Elizabeth Turtle of the University of Arizona, Tucson. "There are linear features, circular features, curvilinear features. These suggest geologic activity on Titan, but we really don't know how to interpret them yet. We've got some exciting work cut out for us." Since entering orbit, Cassini has also provided the first view of a vast swarm of hydrogen molecules surrounding Titan well beyond the top of Titan's atmosphere. Cassini's magnetospheric imaging instrument, first of its kind on any interplanetary mission, provided images of the huge cloud sweeping along with Titan in orbit around Saturn. The cloud is so big that Saturn and its rings would fit within it. "The top of Titan's atmosphere is being bombarded by highly energetic particles in Saturn's radiation belts, and that is knocking away this neutral gas," said Dr. Stamatios Krimigis of Johns Hopkins Applied Physics Laboratory, Laurel, Md., principal investigator for the magnetospheric imager. "In effect, Titan is gradually losing material from the top of its atmosphere, and that material is being dragged around Saturn."The study of Titan, Saturn's largest moon, is one of the major goals of the Cassini-Huygens mission. Titan may preserve in deep-freeze many chemical compounds that preceded life on Earth. Friday's flyby at a closest distance of 539,000 kilometers (210,600 miles) provided Cassini's best look at Titan so far, but over the next four years, the orbiter will execute 45 Titan flybys as close as approximately 950 kilometers (590 miles). This will permit high-resolution mapping of the moon's surface with an imaging radar instrument, which can see through the opaque haze of Titan's upper atmosphere. In January 2005, the Huygens probe that is now attached to Cassini will descend through Titan's atmosphere to the surface. daer m'm, few intresting thing i have searched out about saturn is narrated here.



but in next two years he discovered that smaller stars on both sides disapperared by reducing thier size.and then he wrote  I DO NOT KNOW WHAT TO SAY ABOUT SUCH AMAZING, UNEXPECTED AND NEW THING. THE LACK OF TIME,AND MY VIEW TO UNDERST AND ALSO FEAR OF FAULT MADE ME SHOCKED. in 1656ad when cristian hyugence discovered the real shape of ring he did not dare to declare itand wrote his message in coded formwhich meant THER APPEARS ATHIN AND WIDE RING AROUND IT, WHICH DOES NOT TOUCH IT,AND INTEND TO LEAN TOWARDS SOLAR PATH. MY FEELING ABOUT

RING IS THAT IT IS ACTUALLY AN C MOON OF DISTANT PAST WHICH GOT FRACTIONALISED BECAUSE OF SATURN'S OWN GRAVTATIONAL PULL FORMING MILIONS OF TINY AND DOZENS OF MAJOR MOONS. THE TRUTH /S I HAVE WRITTEN A 70 -80 PAGES DISSERTATION WHICH DEALS MY VIEW WITH NASA'S CONTEXTS.I THOUGHT TO SBMIT IT TO YOU,BUT YOU SEEMS TO BE VERY BUSY. I AM SENDING A PART OF IT THRU PROGRAMME REPORT FORM

REGARDS
RATNESH


News Release: 2004-222 September 9, 2004
Cassini Discovers Ring and One, Possibly Two, Objects at Saturn
Scientists examining Saturn's contorted F ring, which has baffled them since its discovery, have found one small body, possibly two, orbiting in the F ring region, and a ring of material associated with Saturn's moon Atlas.
A small object was discovered moving near the outside edge of the F ring, interior to the orbit of Saturn's moon Pandora. The object was seen by Dr. Carl Murray, imaging team member at Queen Mary, University of London, in images taken on June 21, 2004, just days

before Cassini arrived at Saturn. "I noticed this barely detectable object skirting the outer part of the F ring. It was an incredible privilege to be the first person to spot it," he said. Murray's group at Queen Mary then calculated an orbit for the object.

Scientists are not sure if the object is alone. This is because of results from a search through other images that might capture the object to pin down its orbit. The search by Dr. Joseph Spitale, a planetary scientist working with team leader Dr. Carolyn Porco at the Space Science Institute in Boulder, Colo., revealed something strange. Spitale said, "When I went to look for additional images of this object to refine its orbit, I found that about five hours after first being sighted, it seemed to be orbiting interior to the F ring," said Spitale. "If this is the same object then it has an orbit that crosses the F ring, which makes it a strange object." Because of the puzzling dynamical implications of having a body that crosses the ring, the inner object sighted by Spitale is presently considered a separate object with the temporary designation S/2004 5 4. S4 is roughly the same size as S3. In the process of examining the F ring region, Murray also detected a previously unknown ring, S/2004 1R, associated with Saturn's moon, Atlas. "We knew from Voyager that the region between the main rings and the F ring is dusty, but the role of the moons in this region was a mystery, " said Murray. "It was while studying the F ring in these images that I discovered the faint ring of material. My immediate hunch was that it might be associated with the orbit of one of Saturn's moons, and after some calculation I identified Atlas as the prime suspect." The ring is located 138,000 kilometers (86,000 miles) from the center of Saturn in the orbit of the moon Atlas, between the A ring and the F ring. The width of the ring is estimated at 300 kilometers (190 miles). The ring was first spotted in images taken after orbit insertion on July 1, 2004. There is no way of knowing yet if it extends all the way around the planet.

Searches will continue for further detections of the newfound body or bodies seen in association with the F ring. If the two objects indeed turn out to be a single moon, it will bring the Saturn moon count to 34. The newfound ring adds to the growing number of narrow ringlets around Saturn.

CASS/NI AT SATURN - Present Position

These images show SIMULATED views from the Cassini spacecraft, from several useful points of view a will vary depending on what's happening in the mission. Click on an image to see a full-screen view. Th images are updated as circumstances warrant, at least once a day.

These computer-rendered images were generated by David Seal using his Solar System Simulator. To a

Cassini's position in the sky or detailed orbital elements for Cassini please visit the Solar System Dynan web site and click on the Ephemerides - "Horizons" fink. Follow the instructions provided.

CASSINI AT SATURN - Huygens Mission to Titan

http://saturn.jptnasa.gov/cgi-

bin/gs2.cgi?path =../multimedia/irnages/artwork/ima

ges/image8.jpg&type = image

Artists Concept: Huygens probe at Titan

The Huygens probe will usher in 2005 with its landmark mission at Titan. After a seven-year journey strapped to the side of the Cassini Orbiter, Huygens will be set free on Dec. 25, 2004. The Probe will coast for 21 days en route to Titan.

Probe Separation and Transit to Titan

Prior to the probe's separation from the orbiter, the "coast" timer will be loaded with the precise time necessary to turn on the probe systems (15 minutes before the initial encounter with Titan's atmosphere). Then the probe will separate from the orbiter and coast to Titan for 21 days with no systems active except for its wake-up timer.

Huygens will separate from Cassini at 30 centimeters (about 12 inches) per second and a spin rate of seven revolutions per minute to ensure stability during the coast and entry phase. Five days following the release of the probe, Cassini will perform a deflection maneuver. This will place the orbiter in the proper geometry to collect the data during the probe mission. The probe will continue in this mode until it reaches the top of Titan's atmosphere.

Titan's nitrogen-rich atmosphere extends 10 times further into space than Earth's atmosphere. This means the outer fringes of Titan's atmosphere reach almost 600 kilometers (373 miles) into space. When the probe detects this region of Titan's atmosphere, the sleep timer will go off, awakening the probe's science instruments.

Huygens is equipped with six science instruments designed to study the content and dynamics of Titan's atmosphere and collect data and images on the surface.

http://saturnfpinasa.gov/cgi-bin/gs2.cgi?path = ../muitimed

ia/images/artwork/images/im

age16.jpg&type = image

Artists Concept: Huygens

probe separating to enter

Titan's atmosphere

Descent Through Titan's Atmosphere

Huygens will make a parachute-assisted descent through Titan's atmosphere, collecting data as the parachutes slow the probe from super sonic speeds. Five batteries onboard the probe are sized for a Huygens mission duration of 153 minutes, corresponding to a maximum descent time of 2.5 hours plus at least 3 additional minutes (and possibly a half hour or more) on Titan's surface. These batteries are capable of generating 1800 Watt-hours of electricity.

The probe's radio link will be activated early in the descent phase, and the orbiter will "listen" to the probe for the next three hours, which includes the descent plus 30 minutes after impact. Not long after the end of this three-hour communication window, Cassini's high-gain antenna (HGA) will be turned away from Titan and toward Earth.

http://saturn.j01.nasa.gov/operations/images/probedescentjpg

Huygens Probe Descent Profile

Mission Timeline Facts All times are given in Spacecraft Event Time (SCET), Universal

Coordinated Time (UTC) Probe Release:

December 25, 2004 02:00 UTC

Probe entry at Titan: January 14, 2005 11:04 UTC

Speed at Entry:

6 kilometers per second Impact Speed:

5 meters per second Mission Duration:

2 to 2:30 hours

Altitude of Cassini during the Huygens Mission: 60,000 kilometers

Data rate to Cassini Orbiter:

8 kilobits per second Total Battery Capacity of Probe = 1800 Watt-hours

The peak heat-flux is expected in the altitude range below 350 kilometers (217 miles) down to 220 kilometers (137 miles), where Huygens rapidly decelerates from about 21,600 kilometers (13,424 miles) per hour to 1,440 kilometers (895 miles) per hour in less than two minutes.

At this speed, the parachute deployment sequence initiates, starting with a mortar pulling out a Pilot Parachute which, in turn, pulls away the aft cover and deploys the Main Parachute. After inflation of the 8.3 meters (27.2 feet) diameter main parachute, the front shield is released to fall from the Descent Module. Then, after a 30 second delay built into the sequence to ensure that the shield is sufficiently far away to avoid instrument contamination, the Gas Chromatograph Mass Spectrometer (GCMS) and Aerosol Collector and Pyrolyser (ACP) inlet ports open and the Huygens Atmospheric Structure Instrument (HAM boom deploy. The Descent Imager/Spectral Radiometer (DISR) cover is ejected two minutes later.

The main parachute is sized to pull the Descent Module safely out of the front shield. It is jettisoned after 15 minutes to avoid a protracted descent and a smaller 3-meter (10-foot) diameter parachute is deployed. The descent will last between two and two and half hours.

During its descent, Huygens' camera will capture more than 1,100 images, while the Probe's other five instruments will sample Titan's atmosphere and determine its composition.

Data from Huygens will be relayed to the Cassini Orbiter passing overhead. The data will be stored onboard Cassini's Solid State recorders (SSR) for playback to Earth. Huygens is managed by the European Space Agency. Complete details on the mission objectives and science can be found on the ESA Huygens Site.

Cassini Imaging Diary

Earth-Moon flyby

Asteroid Masursky flyby Jupiter Encounter

Beyond Jupiter

Approach to Saturn

Saturn Arrival

http://ciclops.IpLarizona.edu/index.phphttp://ciclops.Iplarizona.edu/iss/iss.phphttp://ciclops.lpLarizona. edu/sci/indexphphttp://ciclops.IpLarizona.edu/team/iss team.phphttp://ciclops.IpLarizona.edu/nr index .phphttp://ciclops.IpLarizona.edunr index main.php

http://ciclops.l Sep 16, 2004: SPECIAL RELEASE - Saturn's faintly banded atmosphere is delicately

pl.arizona.edu/ colored and its threadbare rings cross their own shadows in this marvelous natural view.php?id _ 3 color view from Cassini.

50

Captain's Log

Star Date: 2004.253

It all seems now like a reverie. Our glistening, golden shrouded electromechanical beast, with seven lonely years and billions of miles behind it, sets its sights on a softly-hued, giant and dream-like planet.

Carrying us into the force-filled province of this spinning globe, it buzzes an ancient, wayward, dirt-ice slip of a body, a relic of the long-gone nebula that birthed the outer planets around the sun.

Weeks later, it and we hasten soundlessly across a vast and solid sea of rippling waves, scalloped shores, and tumbling rivers of boulders, stones, pebbles, all made of ice, gathering as we fly the sun's rays diffusing upward through the scene below.

Hours later we are coursing above a hazy, pale orange planet-sized moon of enormous mystery, whose veiled surface offers, perchance, the markings of a water-less, frozen, but organic-rich early Earth, ane whose atmosphere may host the same molecular arrangements and behavior that ultimately bore life on our blue ocean planet. Blurred visions of dark and fight, a field of bright white clouds aloft, and

occasional sightings of narrow, sinuous patterns, hauntingly like meandering streams, are transmitted on the backs of the few and the lucky electromagnetic beams that have dodged the haze-filled air to be caught by our speeding scopes overhead.

## CHAPTER-7: COLOMBIA-STS107

STS 107 COLUMBIA SPACE SHUTTLE

(ALL DEAD,BUT LIVES BY MAKING HISTORY IN SPACE JOURNEY)

COLUMBIA SPCE SHUTELE, STS -107 ON ITS SIXTEEN DAY JOURNEY TO NEARER SPACE FROM JANUARY 16/2003 TO FEB 1/2003, CONDUCTED ALMOST 80 EXPERIMENTS ON BIOLOGICAL EFFECTS ON CREATURES IN SPACE AND LIFE SUITABILITY THERE ON ITS 28$^{TH}$ JOURNEY INTO SPACE, ONBOARD IT WERE SEVEN EXPERIENCED ASTRONAUTS, WITH A ZEAL TOWARDS LIFE AND MOTTO TO EXPLORE SPACE FOR MANKIND, ON THE RE-

ENTERY OF SPACE SHUTTLE IN TO EARTH'S ATMOSPHERE ,MINUTES AWAY FROM THEIR HOME LOST THEIR LIVES.

MANY OF US WATCHED THE ENTIRE EVENT BUT RARE ONES MAY DARE TO FOLLOW THE FOOTSTEPS OF SPACE SCIENTISTS RICK D HUSBAND, WILLIUM C MCCOOL.MICHEAL P ANDERSON,DAVID M BROWN,KALPNA CHAWALA,LAUREL BLAIR SAL TON CLARK .

Life Facts
Born:
April 16, 1956
Arlington, Va.
Died:
Feb. 1, 2003
during re-entry of Space Shuttle Columbia
Space Agency:
NASA
Astronaut Class:
1996
Missions:
STS-107
Time in Space:
15 days,
22 hours,
20 minutes

Education:
1978, B.S., College of William and Mary.
1982, M.D., Eastern Virginia Medical School.
Military:
U.S. Navy, Captain
Images
http://spaceflight.nasa.gov/gallery/images/shuttle/sts107/html/s107e05155.html Visit the Gallery to see photos of Mission Specialist David Brown and the STS-107 crew.
Videos
STS-107 Tribute
QuickTime
Media Player - 28K/ 56K Real Video - 28K / 56K
Brown's Preflight Profile QuickTime
Media Player - 28K / 56K Real Video - 28K / 56K
Flight Day 4
Flight Day 15
http://spaceflight.nasa.gov/shuttle/archives/sts-
Wake-up Calls
107/memorial/husband.htmlhttp://spaceflightnasa.gov/shuttle
Flight Day 2 - "EMA EMA" (5.2 /archives/sts-
MB) .way file / Net Show /
107/memorial/mccoothtmlhttp://spaceflight.nasa.gov/shuttle/
RealAudio archiyes/sts-

Flight Day 4 - "Cultural Exchange" (4.9 MB) .way file / Net Show / RealAudio

Flight Day 6 - "Texan 60" (3.6 MB)

Flight Day 12 - "When Day is Done" by Django Reinhardt and Stephane Grappelli. (4.4 MB) .way file / Net Show / RealAudio

Flight Day 14 - "I Get Around" by The Beach Boys. (5 MB) .way file / Net Show / RealAudio

Flight Day 96 - "Silver Inches" by

107/memorial/chawla.html http://spaceflight.nasa.gov/shuttle/ archives/sts-107/memorial/anderson.html http://spaceflight.nasa.gov/shuttle/archives/sts-

"I was born April 16, 1956, in Arlington, Va. While growing up he didn't have a specific dream of becoming an astronaut. Although he thought that being an astronaut would be a good job, he didn't think it was possible. "I was a little bit late for Mercury, but I remember Gemini and Apollo quite well in the Sixties, and then Skylab and early shuttle," he said in a preflight interview. "But I absolutely couldn't identify with the people who were astronauts. I thought they were movie stars. And I lust thought I was kind

Life Facts Born:
December 25, 1959
Plattsburgh, N.Y.

Died:
Feb. 1, 2003
during re-entry of Space Shuttle Columbia Space Agency:
NASA
Astronaut Class:
1995
Missions: STS-89
STS-107
Time in Space:
24 days, 18 hours, 7 minutes Education:
1981, BS, University of Washington.
1990, M.S., Creighton University.
Military:
U.S. Air Force, Lieutenant Colonel
Images

Visit the Gallery to see photos of Mission Specialist Michael Anderson and the STS-107 crew. Videos
STS-107 Tribute
QuickTime
Media Player - 28K 56K
Real Video - 28K/ 56K
Preflight Profile
QuickTime
Media Player - 28K/ 56K
Real Video - 28K/ 56K

http://spaceflight.n
107/memorial/husl
e/archives/sts-
107/memorial/mcc
/ archives/sts-
107/memorial/brov
archives/sts-
107/memorial/cha\
archives/sts-
107/memorial/clark
chives/sts-107/mer
always wanted to fl
colonel in the Air Fc
"Along the way, he
daughters and for t
said to them, 'What
for it now."

Life Facts

Born:

March 10, 1961 Ames, Iowa

Died:

Feb. 1, 2003

during re-entry of Space Shuttle

Columbia

Space Agency: NASA

Astronaut Class: 1996

Missions:

STS-107

Time in Space: 15 days,

22 hours,

20 minutes

Education:

1983, B.S., University of

Wisconsin-Madison.

1987, M.D., University of Wisconsin-Madison

Military:

U.S. Navy, Captain Images

Visit the Gallery to see photos of Mission Specialist Laurel Clark and the STS-107 crew.

Videos

STS-107 Tribute

QuickTime

Media Player - 28K / 56K

Real Video - 28K / 56K

Flight Day 3

Flight Day 4

Flight Day 15

Wake-up Calls

Flight Day 5 - 'Amazing Grace" by The Black Watch and the band of the 51st Highland Brigade.

(6.2 MB) .way file / Net Show / RealAudio

Flight Day 13 - "Running to the Light" by Runrig.

(6.6 MB) . way file / Net Show / RealAudio

Flight Day 17 - "Scotland the Brave" by The Black Watch and the band of the 51st Highland Brigade.

(5.7 MB) .wav file / Net Show I RealAudio

Related Links

A Letter to America from the Columbia Crew Families

Clark's Biography

Clark's Preflight Interview Clark's STS-107 Menus

STS-107 Science

*http://spaceflight.nasa.gov/shuttle/archives/sts-107/memorial/husband.htmlhttp://spaceflight.nasa.gov/shuttle/archives/sts-*

*107/memorial/mccool.htmlhttp://spaceflight.nasa.gov/shuttle/archives/sts-107/rnemorial/brown.htmlhttp://spaceflight.nasa.gov/shuttle/archives/sts-107/memorial/chawla.htmlhttp://spaceflight.nasa.gov/shuttle/archives/sts-107/memorial/anderson.htmlhttp://spaceflight.nasa.gov/shuttle/archives/sts-107/memorial/ramon.htmlU.S. Navy Captain Laurel Salton Clark's path to becoming an astronaut evolved over time. Clark said that while growing up she had an interest in the environment and animals.*

*"I was interested in the Moon landings just about the same as everyone else of my generation," she said. "But, I never really thought about being an astronaut or working in space myself. I was very interested in environment and ecosystems and animals."*

*She said her parents were a huge influence on her fife when she was a child. "They always expected the most out of all of us," she said, "and expected us to do our very best."*
*http://spaceflight.nasa.gov/shuttle/crew/intclark.htmIClark*


Life Facts
Born:

Sept. 23, 1961
San Diego, Calif.
Died:
Feb. 1, 2003
during re-entry of Space Shuttle Columbia
Space Agency:
NASA
Astronaut Class:
1996 Missions:
STS-107
Time in Space:
15 days,
22 hours,
20 minutes
Education:
1983, B.S., U.S. Naval Academy.
1985, M.S., University of Maryland.
1992, MS, U.S. Naval Postgraduate School.
Military:
U.S. Navy, Commander
Images
*http://spaceflight.nasa.gov/gallery/*images/shuttle/sts-107/html/s107e05026.html
Visit the Gallery to see photos of Pilot Willie McCool and the STS-107 crew. Videos

STS-107 Tribute
QuickTime
Media Player - 28K / 56K
Real Video - 28K / 56K
McCool's Preflight Profile

***QuickTime***
  *http://spaceflight.nasa.gov/shuttle/archives/sts-*
***Media Player - 28K/ 56K***
  *107/memorial/husband.htmlhttp://spaceflight.nasa.gov/shuttl*
***Real Video - 28K / 56K***  *e/archives/sts-*
***Flight Day 11***
  *107/memorial/brown.htmlhttp://spaceflight.nasa.gov/shuttle/*
***Flight Day 15***  *archives/sts-*
***Wake-up Calls***
  *107/memorial/chawla.htmihttp://spaceflight.nasa.gov/shuttle/*
***Flight Day 3 -*** *"Coming Back to Life"* *archives/sts-*
*by Pink Floyd. **(5.6 MB)** .way file* / *107/memorial/anderson.htmlhttp://spaceflight.nasa.gov/shutt* ***Net Show** / **RealAudio*** *le/archives/sts-*
***Flight Day 5 -*** *"Fake Plastic Trees"* *107/memorial/clark.htmlhttp://spaceflight.nasa.gov/shuttle/ar* *by Radiohead. **(6.2 MB)** . way file* / *chives/sts-107/memorial/ramon.html*Before U.S. Navy

_Net Show_ / _RealAudio_     Commander William McCool began his 16-day scientific Flight Day 10 - "Hotel California" mission, he explained what was most important about the performed by McCool's family. _(4.4_ work he would be doing in space.
_MB) . way file_ / _Net Show_ /
_RealAudio_    "Most of what we're doing is enabling technology for the
Flight Day 15 - "Imagine" by John future," he said. "And the folks who are going to use that
Lennon. _(8.5 MB) . wav file_ / _Net_    technology and then continue the wheels turning are the
_Show_ / _RealAudio_    children today. There's no greater experience, at least in my
Related Links career thus far, than to see the excitement and the eyes fight
_A Letter to America from the_ up when you talk to kids about experiments."
_Columbia Crew Families_    McCool was born in San Diego, Calif:, in 1961. After graduating
_McCool's Biography_ from Coronado High School, Lubbock, Texas, in 1979, McCool
_McCool's Preflight Interview_     went to the U.S. Naval Academy in Annapolis. Md. He

Life Facts
Born:
July 12, 1957
Amarillo, Texas
Died:
Feb. 1, 2003
during re-entry of Space Shuttle Columbia
Space Agency:
NASA
Astronaut Class:
1995 Missions:
STS-96
STS-107
Time in Space:
26 days,
3 hours,
33 minutes
Education:
1980, B.S., Texas Tech University.
1990, M.S., CaL State University, Fresno
Military:
U.S. Air Force, Colonel
Images

Visit the Gallery to see photos of Commander Rick Husband and the STS-107 crew. Videos
STS-107 Tribute
QuickTime Media Player - 28K / 56K Real Video - 28K / 56K
Flight Day 11
Flight Day 14
Flight Day 2 - "America, the Beautiful" by the Texas Elementary Honors Choir, with Husband's daughter, Laura. (7.6 MB) . way file / Net Show / RealAudio
Flight Day 6 - "God of Wonders" by Steve Green. (6.3 MB) . wave file / Net Show / RealAudio

Wake-up Calls
http://spaceflight.nasa.gov/shuttle/archives/sts-107/memorial/mccoolhtml
http://spaceflight.nasa.gov/shuttle/archives/sts-107/memorial/brown.html
http://spaceflight.nasa.gov/shuttle/archives/sts-107/memorial/chawla.html
http://spaceflightnasa.gov/shuttle/archives/sts-107/memorial/anderson.html
http://spaceflight.nasa.gov/shuttle/ar 107/memorial/clark.html

Flight Day 10 - "The Prayer" (4.2chives/sts-107/meinorial/ramon.htmlU.S. Air Force Col. Rick MB) . way file / Net Show / Husband's childhood dream was to become an astronaut. He
RealAudio said that the early human space flight programs -- Mercury,
Flight Day 14 - "Up on the Roof" Gemini and Apollo -- made an impression on him. "...watching by James Taylor.the Moon landings and everything," he said, "it was just so
(7.8 MB) . wav file / Net Show / incredibly adventurous and exciting to me that I just thought,
RealAudio 'There is no doubt in my mind that that's what I want to do
Related Links when I grow up."
A Letter to America from the The Amarillo, Texas, native was born in 1957. Crowing up in
Columbia Crew Families West Texas he developed an interest in flying. "I'd be out in
Husband's Biography my backyard playing," he said in a preflight crew interview.
Husband's Preflight Interview "And, any time I heard any kind of an airplane, you know, it's
Husband's STS-107 Menus like, stop what you're doing and take a look and see, 'Where's
STS-107 Science that airolane? What kind is it? Where is it aoina? How hiah is

Background Information on the Columbia Space Shuttle Mission STS-107
STS-107 Mission Summary
STS-107 Flight: January 16-February 1, 2003
Crew:
Commander Rick D. Husband (second flight), Pilot William C. McCool (first flight),
Payload Specialist Michael P. Anderson (second flight),
Mission Specialist Kalpana Chawla (second flight), Mission Specialist David M. Brown (first flight), Mission Specialist Laurel B. Clark (first flight), Payload Specialist Ilan Ramon, Israel (first flight)
Payload:
First flight of SPACEHAB Research Double Module,˙ Fast Reaction Experiments Enabling Science, Technology, Applications and Research (FREESTAR); first Extended Duration Orbiter (EDO) mission since STS-90. This 16-day mission was dedicated to research in physical, life, and space sciences, conducted in approximately SO separate experiments, comprised of hundreds of samples and test points. The seven astronauts worked 24 hours a day, in two alternating shifts.
First flight:
April 12-14, 1981 (Crew John W. Young and Robert Crippen)
28 flights 1981-2003.
Most recent flight:

STS-109, March 1-12, 2002 Hubble Space Telescope Servicing Mission
Other notable missions:
STS 1 through 5, 1981-1982 first flight of European Space Agency built Spacelab. STS-50, June 25-July 9, 1992, first extended-duration Space Shuttle mission. STS-93, July 1999 placement in orbit of Chandra X-Ray Observatory.
Past mission anomaly:
STS-83, April 4-8, 1997. Mission was cut short by Shuttle managers due to a problem with fuel cell No.

2, which displayed evidence of internal voltage degradation after the launch.
Background Information on the Columbia Space Shuttle Mission STS-107
STS-107 Mission Summary
STS-107 Flight: January 16-February 1, 2003
Crew:
Commander Rick D. Husband (second flight), Pilot William C. McCool (first flight),
Payload Specialist Michael P. Anderson (second flight),
Mission Specialist Kalpana Chawla (second flight),
Mission Specialist David M. Brown (first flight),
Mission Specialist Laurel B. Clark (first flight),
Payload Specialist Ilan Ramon, Israel (first flight)

Payload:

First flight of SPACEHAB Research Double Module; Fast Reaction Experiments Enabling Science, Technology, Applications and Research (FREESTAR); first Extended Duration Orbiter (EDO) mission since STS-90. This 16-day mission was dedicated to research in physical, life, and space sciences, conducted in approximately 80 separate experiments, comprised of hundreds of samples and test points. The seven astronauts worked 24 hours a day, in two alternating shifts.

First flight:

April 12-14, 1981 (Crew John W. Young and Robert Crippen)
28 flights 1981-2003.

Most recent flight:

STS-109, March 1-12, 2002 Hubble Space Telescope Servicing Mission

Other notable missions:

STS 1 through 5, 1981-1982 first flight of European Space Agency built Spacelab. STS-50, June 25-July 9, 1992, first extended-duration Space Shuttle mission. STS-93, July 1999 placement in orbit of Chandra X-Ray Observatory.

Past mission anomaly:

STS-83, April 4-8, 1997. Mission was cut short by Shuttle managers due to a problem with fuel cell No. 2, which displayed evidence of internal voltage degradation after the launch.

STS-107 Shuttle Mission Imagery - Mission Control Center

| http://spaceflight.nas agov/gallery/images/ shuttle/sts-107/html/jsc2003-00010.html | http://spaceflight.nas a.gov/gallery/images/ shuttle/sts-107/html/jsc2003-00056.html | http://spaceflight.nas agov/gallery/images/ shuttle/sts-107/html/j 00058.htm |
|---|---|---|
| http://spaceflight.nas a.gov/gallery/images/ shuttle/sts-107/html/jsc2003- | http://spaceflight.nas a go v/gallery/images/ shuttle/sts-107/html/jsc2003- | http://spaceflight.nas a go v/gallery/images/ shuttle/sts-107/html/j |

Select image for high or low resolution and caption.
Use arrows or page numbers below for more thumbnails.
STS-107 Shuttle Mission Imagery - Flight Day 01
http://spaceflight.nas
agov/gallery/images/
shuttle/sts-
107/htm1/03pd0109.h
tml

http://spaceflightnas
agov/gallery/images/
shuttle/sts-
107/htm1/03pd0113.h
tml

http://spaceflight.nas
agov/gallery/images/
shuttle/sts-
107/htm1/03pd0115.h
tml

http://spaceflight.nas
agov/gallery/images/
shuttle/sts-
107/html/03pd0116.h
tml

STS-107 Shuttle Mission Imagery - Flight Day 02
http://spaceflight.nas
a.gov/gallery/images/
shuttle/sts-
107/html/s107e05001.
html

http://spaceflight.nas
a.gov/gallery/images/
shuttle/sts-
107/html/s107e05002.
html

http://spaceflight.nas
a.gov/gallery/images/
shuttle/sts-
107/html/s107e05003.
html

http://spacellight.nas
a.gov/gallery/images/
shuttle/sts-
107/html/s107e05006.
html

http://spaceflight.nas
agov/gallery/images/
shuttle/sts-

http://spaceflight.nas
agov/gallery/images/
shuttle/sts-

http://spaceflight.nas
agov/gallery/images/
shuttle/sts-

http://spaceflight.nas
agov/gallery/images/
shuttle/sts-

STS-107 Shuttle Mission Imagery - Flight Day 03

http://spaceflight.nas a.gov/gallery/images/ shuttle/sts-107/html/s107e05010.html

http://spaceflight.nas a.gov/gallery/images/ shuttle/sts-107/html/s107e05014.html

http://spaceflight.nas a.gov/gallery/images/ shuttle/sts-107/html/s107e05018.html

http://spaceflight.nas a.gov/gallery/images/ shuttle/sts-107/html/s107e05020.html

http://spaceflight.nas a.gov/gallery/images/ shuttle/sts-107/html/s107e05059.html

http://spaceflight.nas a.gov/gallery/images/ shuttle/sts-107/html/s107e05054.html

http://spaceflight.nas a.gov/gallery/images/ shuttle/sts-107/html/s107e05060.html

http://spaceflight.nas a.gov/gallery/images/ shuttle/sts-107/html/s107e05070.html

http://spaceflight.nas a.gov/gallery/images/ shuttle/sts-107/html/s107e05096.html

http://spaceflight.nas a.gov/gallery/images/ shuttle/sts-107/html/s107e05098.html

http://spaceflight.nas a.gov/gallery/images/ shuttle/sts-107/html/s107e05100.html

http://spaceflight.nas a.gov/gallery/images/ shuttle/sts-107/html/s107e05103.html

## *STS-107 Shuttle Mission Imagery - Flight Day 04*

*http://spaceflight.nas  http://spaceflightnas*
*http://spaceflight.nas  http://spaceflightnas*
*a.gov/gallery/images/  agov/gallery/images/*
*a.gov/gallery/images/  a.gov/gallery/images/*
*shuttle/sts-  shuttle/sts-  shuttle/sts-*
*shuttle/sts-*
*107/html/s107e05135.  107/html/s107e05152.*
*107/html/s107e05155.  107/html/s107e05167.*
*html  html  html  html*
**STS-107 Shuttle Mission Imagery - Flight Day 05**
*http://spaceflight.nasagovhttp://spaceflight.nasa.go*
*http://spaceflight.nasa.go /gallery/images/shuttle/sts*
*v/gallery/images/shuttle/s  v/gallery/images/shuttle/s*
*ts-*
*-107/html/s107e05208.html[ts-]*
*107/html/s107e05214.html107/html/s107e05220.ht*
*m4*

**STS-107 Shuttle Mission Imagery - Flight Day 06** *http://spac*
*http://spaceflight.nas  http://spaceflight.nas  a.gov/galle*
*a.gov/gallery/images/ a.gov/gallery/images/ shuttle/sts-  shu*
*shuttle/sts-*
*107/html/s107e05405.  107/html/s107e05407.  107/html/s107e0.*
*html  html  html*
*http://spaceflight.nas  http://spaceflight.nas*
*agov/gallery/images/ a go v/gallery/images/*
*shuttle/sts-  shuttle/sts-*
**STS-107 Shuttle Mission Imagery - Flight Day 07**

http://spaceflight.nas a.gov/gallery/images/ shuttle/sts-107/html/s107e05344.html

http://spaceflight.nas agov/gallery/images/ shuttle/sts-107/html/s107e05354.html

http://spaceflight.nas http://spaceflight.nas a.gov/gallery/images/ shuttle/sts-107/html/s107e05353.html

http://spaceflight.nas agov/gallery/images/ shuttle/sts-107/html/s107e05359.html

http://spaceflight.nas a.gov/gallery/images/ shuttle/sts-107/html/s107e05484.html

http://spaceflight.nas agov/gallery/images/ shuttle/sts-107/html/s107e05485.html

**STS-107 Shuttle Mission Imagery - Flight Day 08**

http://spaceflight.nasa.go ov/gallery/images/shuttle X07/htm//s107e05369.htm/ 107/html/s107e05369.html

http://spaceflight.nasa.gov v/gallery/images/shuttle/s /sts-107/html/s107e05378.ht ml

http://spaceflight nasa.g /gallery/images/shuttle/sts

**STS-107 Shuttle Mission Imagery - Flight Day 09**

http://spaceflight.nasa.go v/gallery/images/shuttle/s X07/html/s107e05682.html

http://spaceflight.nasa.go v/gallery/images/shuttle/s

http://spaceflight.nasa.gov /gallery/images/shuttle/sts

*107/html/s107e05682.html107/html/s107e05684.html*

***STS-107 Shuttle Mission Imagery - Flight Day 10***
*http://spaceflight.nasa.gov/gallery/i*
*http://spaceflight.nasa.gov/gallery*
*mages/shuttle/sts- /images/shuttle/sts-107/html/s107e05537.html*
*107/html/s107e05685.html*

Page (@

5- 107/html/sts107- 402-012.html http://spaceflight.nasa.gov/gallery/ images/shuttle/st 5-107/html/sts107- 401-006.html http://spaceflight.nasa.gov/gallery/ images/shuttle/st s-107/html/sts107- 400-010.html http://spaceflight.nasa.gov/gallery/ images/shuttle/st s-

107/html/sts107-400-004.html http://spaceflight.nasa.gov/gallery/ images/shuttle/st 5-107/html/sts107- 399-036.html http://spaceflight.nasa.gov/gallery/ images/shuttle/st s-107/html/sts107-399-006.html

http ✓ /spaceflight .nasa.gov/gallery/ images/shuttle/st 5-107/html/sts107- 399-004.html
glovebox on the middeck of the Space Shuttle Columbia. EDITOR'S NOTE: On February 1, 2003, the seven...

JPEG Graphics Format
STS107-401-006 (16 January - 1 February 2003) --- Astronaut Rick D. Husband, STS-107 mission commander, is photographed near the control panels and windows on the aft flight deck of the Space Shuttle Columbia. EDITOR'S NOTE: On February 1, 2003, the seven crewmembers were lost with the...

JPEG Graphics Format
STS107-400-010 (16 January - 1 February 20031--- This view featuring Saudi Arabia, the Red Sea, Egypt, and the Nile River was photographed by an STS-107 crewmember onboard the Space Shuttle Columbia. EDITOR'S NOTE: On February 1, 2003, the seven crewmembers were lost with the Space...

JPEG Graphics Format
STS107-400-004 (16 January - 1 February 2003) --- This Earth view featuring the Sinai Peninsula, Red Sea, Egypt, Nile River, and Mediterranean was

photographed by an STS-107 crewmember onboard the Space Shuttle Columbia. EDITOR'S NOTE: On February 1, 2003, the seven crewmembers were lost with the...

JPEC Graphics Format

STS107-399-036 (16 January - 1 February 20031--- A "fish-eye" lens on a 35mm camera records astronaut David M. Brown, STS-107 mission specialist, as he peers through a portal window located on the overhead bulkhead in the SPACEHAB Research Double Module (RDM) aboard the Space Shuttle...

JPEG Graphics Format

STS107-399-006 (16 January - 1 February 2003) --- Astronaut Rick D. Husband, STS-107 mission commander, holds a checklist as he speaks into a portable microphone on the aft flight deck of the Space Shuttle Columbia EDITOR'S NOTE: On February 1, 2003, the seven crewmembers were lost...

JPEC Graphics Format

STS107-399-004 (16 January - 1 February 2003) --- Astronaut Rick D. Husband, STS-107 mission commander, is pictured on the aft flight deck of the Space Shuttle Columbia. EDITOR'S NOTE: On February 1, 2003, the seven crewmembers were lost with the Space Shuttle Columbia over North Texas....

JPEG Graphics Format

Search: Shuttle
Keywords:
220 Matches found Page 7 of 22
http://spaceflight .nasa.gov/gallery/ images/shuttle/sts-
107/html/s107e05 16Zhtml
http://spaceflight .nasa.gov/gallery/ images/shuttle/sts-
107/html/s107e05 155.html
STS107-E-05167 (19 January 2003) --- Astronaut Laurel B. Clark, STS-107 mission specialist, takes a brief break from science research on the mid deck of the Space Shuttle Columbia. A sleep area, composed of adjacent bunk beds, is just out of frame at left.
JPEG Graphics Format
STS107-E-05155 (19 January 2003) --- Astronaut David M. Brown, STS-107 mission specialist, stabilizes the digital video camera (DVCAM) on a wall mount stand prior to a "shoot" in the SPACEHAB Research Double Module aboard the Space Shuttle Columbia.
JPEG Graphics Format


http://spaceflight .nasa.gov/gallery/ images/shuttle/sts-
107/html/s107e05 152.htrni

http://spaceflight.nasa.gov/gallery/ images/shuttle/sts-107/html/s107e05 135.html
http://spaceflight.nasa.gov/gallery/ images/shuttle/sts-107/html/s107e05 103.html
http://spaceflight.nasa.gov/gallery/ images/shuttle/sts-107/html/s107e05 100.html
http://spaceflight.nasa.gov/gallery/ images/shuttle/sts-107/html/s107e05 098.html
http://spaceflight.nasa.gov/gallery/ images/shuttle/sts-107/html/s107e05 096.html
http://spaceflight.nasa.gov/gallery/ images/shuttle/sts-107/html/s107e05 070.html
http://spaceflight.nasa.gov/gallery/ images/shuttle/sts-107/html/s107e05 060.html
Search.Shu
KeYwOr *
gVggir:ftes'fauq Page 11 Of 22
STS107-E-05152 (19 January 2003) --- Astronaut Kalpana Chawla, STS-107 mission specialist, keeps up with the brisk stream of science data in the SPACEHAB Research Double Module aboard the Space Shuttle Columbia.

JPEG Graphics Format
STS107-E-05135 (19 January 2003) --- The tunnel linking the SPACEHAB Research Double Module to the Space Shuttle Columbia's crew cabin provides an interesting point of view in this scene of astronaut Michael P. Anderson, mission specialist, performing work in SPACEHAB.
JPEG Graphics Format
STS107-E-5103 (18 January 2002) --- The tunnel linking the SPACEHAB Research Double Module to the Space Shuttle Columbia's crew cabin provides an interesting point of view in this scene of astronaut Kalpana Chawla, mission specialist, performing work in SPACEHAB. JPEG Graphics Format
STS107-E-5100 (18 January 2003) --- Astronaut Kalpana Chawla, mission specialist, works at an experiment rack in the SPACEHAB Research Double Module onboard the Space Shuttle Columbia.
JPEC Graphics Format
STS107-E-5098 (18 January 2003) --- Astronaut Rick D. Husband, mission commander, works at an experiment rack in the SPACEHAB Research Double Module onboard the Space Shuttle Columbia.
JPEG Graphics Format
STS107-E-05096 (18 January 2003) --- Astronauts Rick D. Husband and Kalpana Chawla, mission commander and mission specialist, respectively,

share an experiment-monitoring chore on the SPACEHAB Research Double Module onboard the Space Shuttle Columbia.
JPEG Graphics Format
STS107-E-05070 (18 January 2003) --- The bright sun dissects the airglow above Earth's horizon in this digital still camera's view photographed from the Space Shuttle Columbia
JPEG Graphics Format
5T5107-E-05060 (18 January 2003) --- After pinpointing a photographic target of opportunity on Earth, astronaut David M. Brown, STS-107 mission specialist, takes a photograph with a still camera.
JPEG Graphics Format

http://spacefli JSC2003-E-06822 (16 January 2003) -- These are two composite images ght.nasa.gov/g of Columbia during ascent on STS-107. The images are both derived allery/images/s from an average of 17 video fields totalling about one-quarter of *efforts since the loss of Columbia Feb. 1.*

*The Columbia Accident Investigation Board recently determined the material was not relevant to their investigation. The imagery documents the STS-107 mission from the crew's perspective. The imagery includes almost 10 hours of recovered video and 92 photographs. It includes in-cabin, Earth observation and experiment-related imagery. The Shuttle carried 337 videotapes, but only 28 were found with some recoverable footage. The mission carried 137 rolls of film, but only 21 were found containing recoverable photographs.*

*The imagery is among the more than 84,000 pieces of debris recovered. The debris weighs*

so, a quick logging into activities of iss...exped 9 and active response of nasa scientists and reporter of bbc and others saved many lives
but it was an international space station which told about hurricanes in many hours advance and saved thousands of lives across usa and world.
ratnesh dwivedi
sep. 16/2004 4:30 indian time
ratniesh dwiviedi in consent with white house government.
ratniesh dwiviedi is an international journalist and is approachable thru e-mails and phones and e-fax
ratnieshdwiviedigmilitary.com
ph.. # 91 08025287140 e-fax...1--214--481—1501
ratnieshclwiviedigmilitary.com
ratniesh007@military.con; e-fax---1-214-481-1501
s/o thakur pd. dubey vill--shahnewaj pur p.o--darshan nagar faizabad/.u.p./ india


Previous / Next Bottom of Form Kennedy Space Center is closed until Sunday, September 12 due to damages caused by Hurricane Frances. I have intermittent access to email but cannot reply. Access to attachments is also very limited. If hurricane Ivan does not hit us this weekend, I should be back at the office on Monday.

This is how international space station ---expedition -9 is working and informed about ivan, Jeanne and frances frances hurricanes. this was give a shape of an article by me, and transferred to white house and various centers of nasa...

dear bbc people and my praposed nasa advisers,

news of Ivan and its disasterous impacte on southern coast of america is tragic. i got to know about Ivan thru my nasa page.

an international space station expedition -9 is orbiting around earth. its a manned mission and few astronauts are constantly busy telling about bad weather of earth. this is how i came to know about the weather,ivan,frances and zeanae, which is going to hit american and islands in pacific ocean.

this is where a international space station gives a cutting edge to a general space mission. the auto responder o my carlos calle was a quicker for me and i jumped into my

pages of nasa.soon i find the space station zooming in to african contenent and few hours later clicking extensivly on Frances and ivvan.

now the crew members are relaxed a little and they have recorded a five hours twenty nine minutes of space walk. commander gennedy padalka and nasa science officer mike fincke wrapped up a succes full space walk at 6:03 edt and for five hrs and 21 minutes.

they replaced and installed a new equippment outside orbiting space stations so far man has made 250 space walks out of whic 149 has been successfully attempted by nasa astronauts.

padalka wearing red striped orlan space suites and nasa's fincke wearing blue stripes begin thier walk at 12:43 edt.

the first talk they finished was to replace a pump panel atop zarya module. this monitors coolent levels .higher the temprature greater to coolent floe thru radiater.next space walkers installed fairleads, called pigtails because of thier circular shape.

task finished before orbital sunset which which comes evey almost four hours, then both remained still as motionlesstest.

there is report that expedition 10 members would be flown on oct 9 and will return along with expedition - 9 members from orbiting space station.expeditin B members were allowed to talk to thier family members in jan--04.

hence forth iss exped--9 and its crew members played heroic role along with nasa scientists likk carlos calle to tell it hours and dayes in advance about massive hit of francess, ivvan and zeanne hurricanes .i took immediate notice and put it on high level of alert .pssed it across bbc world , white house and across globe.

these hurricanes causes severe casualities and weather disturbances in usa and neighbouring islands uncomparable to any part of world.

Discovery Launch

Pad: TBD

Launch: Planning window no earlier than March 6 to April 18, 2005

Window: TED

Docking:

**KSC Prepares for Hurricane Jeanne**

Kennedy Space Center, Fla., is preparing for possible effects from Hurricane Jeanne this weekend. KSC was closed Friday to everyone but essential personnel needed to prepare for the storm.

If Jeanne makes landfall in Florida, it will be the fourth hurricane to hit the state in less than two months. Kennedy was closed 11 days earlier this month due to Hurricane Frances.

The three Space Shuttle orbiters, the Shuttle launch pads and Station flight hardware at KSC sustained no damage during

| | |
|---|---|
| TBD | the previous storms. However, the Vehicle Assembly Building, |
| EVAs: | the Thermal Protection Facility and the Processing Control |
| 3 spacewalks | Center received significant storm damage. |
| Undocking: | For the latest forecast information on Jeanne, please visit the |
| TED | National Hurricane Center's Web site. |
| Landing: | Check out the Gallery to see International Space Station |
| TED | imagery of Hurricanes Ivan, Charley and Frances. |
| Duration: | Discovery Milestones Set Stage for Return to Flight |
| TED | The Readers' Room |

Orbital EVA Training
Insertion

http://spaceflight.nasa.gov/gallery/images/shuttle/sts-

| | |
|---|---|
| Altitude: | 114/html/jsc2003e02626.html |
| 122 nautical | Mission Specialist Stephen Robinson suits up to train for a |
| miles | space walk. |

Orbit STS-114 to Demonstrate Repair Techniques, Deliver Equipment
to Space Station

51.60° The STS-114 crewmembers will deliver supplies to the

Bottom of Form  International Space Station, but the major focus of their

mission will be testing and evaluating new Space Shuttle flight safety, which includes new inspection and repair techniques. STS-114 is classified as Logistics Flight 1. Among the Station-related activities are delivering new supplies and replacing one of the orbital outpost's Control Moment Gyroscopes (CMGs). STS-114 will also carry a Raffaello Multi-Purpose Logistics Module and the External Stowage Platform-2.

The crew is slated to conduct at least three spacewalks while at the MS. The first spacewalk will demonstrate repair techniques of the Shuttle's Thermal Protection System. During the second, the spacewalkers will replace the failed CMG with one delivered by the Shuttle. On the third, they will install the External Stowage Platform.

The auto responder of dr carlos calle about ivan and Frances hurricanes

Printable Version                    Flag this m

From: "Calle, Carlos I" <Carlos.I.Calle@nasagov>
!Save Address] (Block Sender! !This Is Spaml Top of Form

To: "ratnesh dwivedi" <ratneshdwivedi@indiainfo.coin>

CC: ...................

*Support systems Space Product Development Biological processes Crystallography and into fife sciences, combustion research, material sciences and fluid physics. Most of the life science experiments included specimen samples that were lost in the accident, but many of the other experiment results were downlinked during tholecular Structure Among the scientific successes of STS-107 were the six*

*Drug Delivery experiments of the <u>Space Technology and Research Students,</u>*

*Physical processes or STARS, program. The student-designed experiments*

*Technology Development downlinked video and data to the researchers every day, and*

*Attitude Control  an estimated 70 percent of the scientific objectives were*

*Communications  achieved.*

*Educational*

*Satellite sensor calibration*

*When Columbia was on its way,1 was busy with my work on space science,alos into writing e-mail dispatches to president bush on iraq policies and science operations.after tragedy I wrote this letter to president bush on his white house e-mail address.below is a repeat copy of similar letter submitted Copresident bush and my friends who helped me in this long research work on space science ,its physics,biology and dynamics, wholely called the "cosmic mask".*

*Home > News & Policies http://wwwwhitehousagov/news/releases/2003/02/print/20030204-1.html*
*For Immediate Release*

*Office of the Press Secretary*

*February 4, 2003*

*END 12:44 P.M. CST*

*A dedication to India born astronaut kalpana chawala b,by nasa 07.26.04*

*But this is how a scientist at national aeronautics and space administration is taught to never give up. columbia's no safe landing only enhanced the will and desire of whitehouse,and big family of nasa for more space missions.*

*That is how international space stationfissl and expedition 9 is telling all of us.the information about ivan,jeannae and Frances hurricanes which might have destroyed Florida centers of nasa was grabbed by investigators at nasa.kenedy space center scientist in space physics department, dr. carlos.i.calle put an autoresponder in his e-mail box,before leaving for vacation. this info was picked up by me and transferred to bbc world wide .next day information was telecasted on its television. this surly saved life of thousands of peopladr.carlos is my proposed adviser on my associateship programme.*

*Imagery*

*In preparation for the arrival of Hurricane Jeanne, worke Reusable Launch Vehicle Hangar unroll long pieces of place on shelves holding Thermal Protection System Facilit equipment. The TPSF suffered extensive damage from I Frances, causing the relocation of equipment and materie*

00 pounds, about 38 percent of the dry weight of Columbia. More than 30,000 people assisted in the search conducted through the combined efforts of NASA, FEMA, EPA, the U.S. and Texas Forest Services. The Columbia Recovery Office at the Johnson Space Center (JSCI was established to continue accepting calls about debris, since the formal search was completed in April. The toll free number to report debris is: 1/866/446-6603.
Selected scenes and photographs will be broadcast on NASA Television today at 12:15 p.m. EDT. News media may obtain the video and photos in their entirety by calling the JSC Media Resource Center

at: 281/483-4231. NASA Television is broadcast on AMC-2, transponder 9C, C-Band, located at 85 degrees West longitude. The frequency is 3880.0 MHz Polarization is vertical and audio is monaural at 6.8 MHz. Information about NASA and the Columbia accident investigation is on the Internet at:
Shuttle Left Wing Cutaway Diagrams
These detailed views represent a space shuttle left wing with Reinforced
Carbon-Carbon, or RCC, panels with only those panels numbered 1 through 10,
16 and 17 shown. Each wing's leading edge has 22 RCC panels.
View 1 Medium Jpeg (612 Kb)
View 1 Large Jpeg (1.1 Mb)
View 2 Medium Jpeg (612 Kb)
View 2 Large Jpeg (1.1 Mb)
View 3 Medium Jpeg (612 Kb)
View 3 Large Jpeg (1.1 Mb) 4- View 4 Medium Jpeg (612 Kb)
View 4 Large Jpeg (1.1 Mb)
View 5 Medium Jpeg (612 Kb)
View 5 Large Jpeg Mb)
STS-107: Home / The Crew / Cargo / Timeline / EVA / Investigation

'Grace,' a 64-processor Origin 2000 supercomputer named after Grace Hopper, a pioneer in computer science

'Turing,' a 24-processor SGI Origin 2000 supercomputer named for Alan Turing, a mathematician and early computer pioneer

Image Right: G. Scott Hubbard (second from left) poses with, left to right, Walt Brooks, Anthony Robbins (SGI), and Tom Edwards at the 'Kalpana' supercornputer dedication.

"With the addition of the SGI Altix system, NASA's high-end computing testbed activities in support of the agency's science and engineering missions are greatly enhanced," said Walt Brooks, chief of the NASA Advanced Supercomputing (NAS) Division at NASA Ames. "Thanks to its outstanding performance capabilities, this supercomputer is helping NASA achieve breakthrough results to meet major challenges in climate and ocean modeling and aerospace vehicle design," Brooks added.

But this is how a scientist at national aeronautics and space administration is taught to never give up.columbia's no safe landing only enhanced the will and desire of whitehouse,and big family of nasa for more space missions.

\ That is how international space station(iss) and expedition 9 is telling all of us. the information about ivan,jeannae and fiances hurricanes which might have destroyed florida centers of nasa was grabbed by investigators at nasakenedy space center scientist in space physics department, dr. carlos. Lcalle put an autoresponder in his e-mail box,before leaving for vacation-this info was picked up by me and transferred to bbc world wide .next day information was telecasted on its people. dr.carlos is my proposed adviser on my associate ship programme.

3000 supercomputer after Kalpana Chawla," said Ames Center Director G. Scott Hubbard. "She was not only a member of the NASA family, but also a special member of our own Ames family. We all miss her and her many contributions to the agency."

At Ames, Chawla had the challenging task of computing the airflow surrounding a jet-supported, delta-wing aircraft during landing. During an interview in 1995, Chawla predicted that her exposure to a wide variety of computer systems at Ames would be especially useful to her as an astronaut.

Of the dozens of experiments successfully conducted by the Columbia crew, Chawla's favorite was the Israeli Mediterranean Dust Experiment, which involved pointing a camera at Earth to study the effects of dust on weather and the environment.

"Fittingly, the SGI Altix 3000 supercomputer that will be named 'Kalpana' is being used to develop substantially more capable simulation models to better assess the evolution and behavior of the Earth's climate system," said Ghassem Asrar, NASA's Deputy Associate Administrator for Earth Science.

The new supercomputer is being used for a group effort by NASA Headquarters, NASA's Jet Propulsion Laboratory (1PL), Pasadena, Calif, NASA Ames and NASA Goddard Space Flight Center, Greenbelt, Md., to deliver high-resolution ocean analysis in the framework of the ECCO (Estimating the Circulation and Climate of the Ocean) Consortium, which involves JPL, the Massachusetts Institute of Technology,

Cambridge, Mass., and the Scripps Institute of Oceanography, La Jolla, Calif

Image Left: NASA Ames Center Director G. Scott Hubbard unveils the SGI Altix 3000 supercomputer named for Kalpana 'KC' Chawla.

Naming the new supercomputer 'Kalpana' follows a long tradition at the research center of naming its new supercomputers after pioneers in the supercomputer industry or individuals who have significantly contributed to research at Ames.

The following are the Ames Research Center's named supercomputers:

1⁻ 'Chapman,' an SGI Origin 3000, 1,024-processor single-image, shared memory system named after Dr. Dean Chapman, a former director of astronautics at Ames who developed heat protection systems for the space shuttle

1⁻ 'Lomax,' a 512-processor SGI Origin 2000 supercomputer named after Dr. Harvard Lomax, a pioneer in computational fluid dynamics who also worked at Ames

11⁻ 'Steger,' a 128-processor Origin 2800 supercomputer named after Joseph Steger, whose work in computational technology revolutionized the use of computers to solve complex aerospace problems

*/17 'Lou,' the main production storage system at the NASA Advanced Supercomputing Division*

The Columbia's pilot was Commander Willie McCool, whom friends knew as the most steady and dependable of men. In Lubbock tod; thinking back to the Eagle Scout who became a distinguished Naval officer and a fearless test pilot. One friend remembers Willie this v was blessed, and we were blessed to know him."

Our whole nation was blessed to have such men and women serving in our space program. Their loss is deeply felt, especially in this pl. so many ofyou called them friends. The people ofNASA are being tested once again. In your grief; you are responding as your friends have wished -- with fops, professionalism, and unbroken faith in the mission of this agency.

Captain Brown was correct: America's space program will go on.

This cause of exploration and discovery is not an option we choose; it is a desire written in the human heart. We are that part of creatior seeks to understand all creation. We find the best among us, send them forth into unmapped darkness, and pray they will return. They4 peace for all mankind, and all mankind is in their debt.

Yet, some explorers do not return. And the loss settles unfairly on a few. The families here today shared in the courage of those they lov now they must face life and grief without them. The 'sorrow is lonely; but you are not alone. In time, you will find comfort and the grace you through. And in God's own time, we can pray that the day ofyour reunion will come.

And to the children who miss your Mom or Dad so much today, you need to know, they love you, and that love will always be with you. were proud of you. And you can be proud of them for the rest of your life.

The final days of their own lives were spent looking down upon this Earth. And now, on every continent, in every land they could see, 11 of these astronauts are known and remembered. They will always have an honored place in the memory of this country. And today I ofl respect and gratitude of the people of the United States.

May God bless you all.

END 12:44 P.M. CST

PRESIDENT BUSH ATTEND MEMORIAL SERVICE FOR COLUMBIA ASTONAUTS

REMARK BY THE PRESIDENT AT THE MEMORIAL SERVICE IN HONOR OF THE STS 107 CREW SPACE COLUBIA LYNDON

B.JOHNSON SPACE CENTER

HONSTON, TEXAS

12:35 PM CST

THE PRESIDENT THEIR MISION WAS ALMOST COMPLETE AND WE LOST THEM SO CLOSE TO HOME THEMEN AND WOMEN OF THE COLUMBIA HA JOURNEYED MORE THAN 6MILLION MILES AND WERE MINUTES AWAY FROM ARRIVAL AND REVNION.

THE LOSS WAS SUDDEN AND TETRIBLE AND FOR THEIR FAMILIES THE GRIEF IS HEAVY OUR NATION SHARES IN YOUR SORROW AND IN YOUR PRIDE AND TOD REMMBER NOT ONLY ONE MOMENT OF TRAGEDY BUT SEVEN

LIVES OF GREAT PURPOSE AND ACHIEVEMENT.

TO LEAVE BEHIND EARTH AND AIR AND GRAVITY IS AN ANCIENT DREAM OF HUMANITY FOR THESE SEVEN IT WAS A DREAM FULFILLED EACH OF THESE ASTR THE DATING AND DISCIPLINE REQUIRED OF THEIR CALLING EACH OF THEM KNEW THAT GREAT ENDAVOTS ARE INSEPARADLE FROM GREAT RISK AND EACH ACCEPTED THOSE RISK WILLINGLY EVEN JOYFULLY IN THE CAUSE OF DISCOVERY.

www.ingramcontent.com/pod-product-compliance
Lightning Source LLC
Chambersburg PA
CBHW050159230526
45470CB00001B/164